Geobotany Studies

Basics, Methods and Case Studies

The series includes outstanding monographs and collections of papers on a range of topics in the following fields: Phytogeography, Phytosociology, Plant Community Ecology, Biocoenology, Vegetation Science, Eco-informatics, Landscape Ecology, Vegetation Mapping, Plant Conservation Biology, and Plant Diversity. Contributions should reflect the latest theoretical and methodological developments or present new applications on large spatial or temporal scales that will reinforce our understanding of ecological processes acting at the phytocoenosis and vegetation landscape level. Case studies based on large data sets are also considered, provided they support habitat classification refinement, plant diversity conservation or vegetation change prediction. ***Geobotany Studies: Basics, Methods and Case Studies*** is the successor to *Braun-Blanquetia*, a journal published by the University of Camerino from 1984 to 2011 in cooperation with the *Station Internationale de Phytosociologie* (Bailleul, France) and the *Dipartimento di Botanica ed Ecologia (*Università di Camerino, Italy) and under the aegis of the *Société Amicale Francophone de Phytosociologie*, the *Société Française de Phytosociologie*, the *Rheinold-Tüxen-Gesellschaft* and the *Eastern Alpine and Dinaric Society for Vegetation Ecology*. This series promotes the expansion, evolution, and application of the invaluable scientific legacy of the Braun-Blanquet school.

More information about this series at http://www.springer.com/series/10526

Mohammed Sghir Taleb • Mohamed Fennane

Vascular Plant Communities of Morocco

Phytosociology, Ecology and Geography

Springer

Mohammed Sghir Taleb
Department of Botany, Institut Scientifique
Mohammed V University in Rabat
Rabat, Morocco

Mohamed Fennane
Department of Botany, Institut Scientifique
Mohammed V University in Rabat
Rabat, Morocco

ISSN 2198-2562 ISSN 2198-2570 (electronic)
Geobotany Studies
Basics, Methods and Case Studies
ISBN 978-3-319-93703-8 ISBN 978-3-319-93704-5 (eBook)
https://doi.org/10.1007/978-3-319-93704-5

Library of Congress Control Number: 2018949066

This Springer imprint is published by the registered company Springer Nature Switzerland AG
The registered company address is: Gewerbestrasse 11, 6330 Cham, Switzerland

Introduction

Phytosociological research in Morocco began to get great importance in the middle of the last century. Since then, numerous studies (publications, thesis, etc.) have been carried out. Thus, during the 1950s, important works were carried out on the Atlas mountains (Quézel 1952, 1957; Nègre 1952, 1961), Western arid Morocco (Nègre 1956, 1959) and the Saharan-Moroccan confines (Lemée 1953). After a phase of "stagnation", which was rather long, Moroccan phytosociology was going to be very dynamic, particularly between 1978 and 1996, thanks to Professor Pierre Quézel and his pupils and collaborators. All regions of the country have been explored and studied, including forests, preforests and matorrals (Quézel et al. 1981–1995; Barbéro et al. 1982; Benabid 1982), the cedar forests (Mhirit 1982, 1987), the tetraclineria (Benabid 1976; Fennane 1987), the arganeria (Peltier 1982), the lagoon and estuarine ecosystems (Bendanoun 1991), etc.

More recently, the national phytosociological scheme has been enriched by the contribution of several authors on somewhat peculiar ecosystems, in particular the psammophilic groups (Gehu and Biondi 1996), the saxicoles (Deil and Galan de Mera 1996), the steppes and pre-steppes of mountains (Taleb and Fennane 2003, 2010), wetlands (Molina et al. 2009), etc.

Nowadays, it could be confirmed that the major phytosociological features of the country's main vegetation structures are known, but they still need to be refined and the gaps are unfortunately still large. This work is in this way hoping to help researchers and attract their attention to plant communities, ecosystems and regions that are little known or have not been studied so far.

Methodology

For more than 5 years, we have been continuously working to gather the existing phytosociological information on Morocco in a database under the Access software. All the syntaxonomic units (classes, orders, alliances and associations) are recorded. According to the available data, for each association or groupement, the following information is stored: scientific names of syntaxa, synonyms, characteristic species, vegetation cover, altitude, soil/substrate, topography, exposure, geographical distribution, bioclimate, vegetation belt, sub-associations and possibly particular remarks.

This database is mainly filled by bibliographic data, after being compiled, analysed and structured. It is the main support for the preparation of our present work synthesis. Obviously, not all the content of the database is included here; only the following important information is given:

- Scientific name of syntaxa, following the nomenclature adopted in Europe (Mucina et al. 2016) and/or in Iberian Peninsula (Rivas-Martinez et al. 2002)
- Vegetation belts, sensu Achhal et al. (1980): Inframediterranean, Thermomediterranean, Mesomediterranean, Superior Mediterranean/ Supramediterranean, Mountain mediterranean, Oromediterranean
- Bioclimate, sensu Emberger (1939) and Sauvage (1963): Saharan, Arid, Semi-arid, Sub-humid, Humid, Hyper-humid; cold, cool, temperate or warm winter
- Altitudes
- Substrates and soils
- Geographical distribution: see Map 1
- Particular remarks about ecology, floristic, dynamic . . .

The current state of knowledge indicates that 670 associations (communities or groupements) have been identified and studied in Morocco, spread over 97 alliances, 66 orders and 44 classes. These associations represent 8 groups of vegetal formations (Table 1). For some syntaxa (association, alliance or even order), the phytosociological position may have changed according to advanced research. We have not been able to take into account all the modifications; we hope doing this, in detail, in a critical revision of the whole phytosociological units of Morocco.

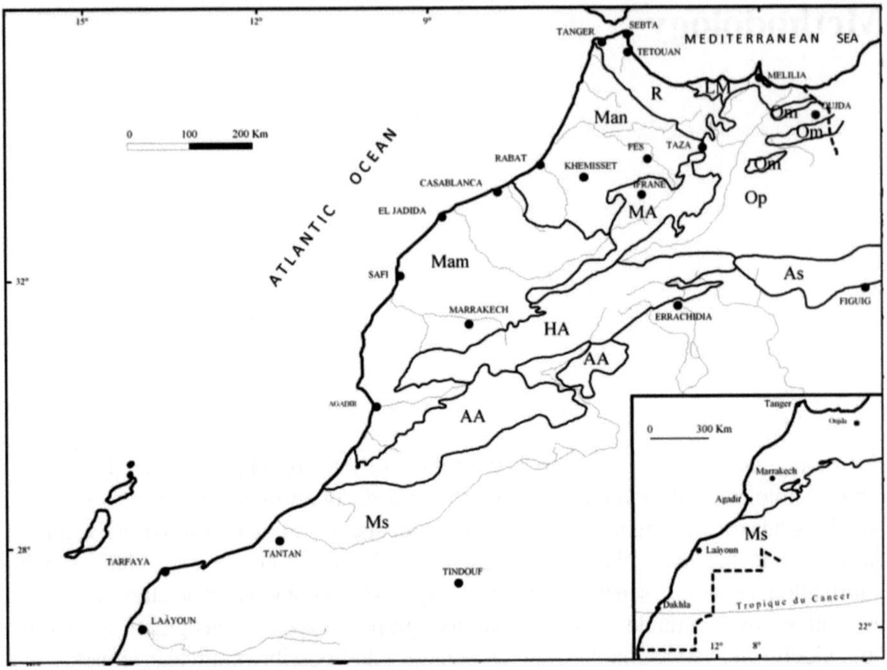

Map 1 Geographic divisions of Morocco. *Ms* Saharan Morocco, *As* Saharan Atlas, *AA* Anti-Atlas, *HA* High-Atlas, *MA* Middle-Atlas, *Mam* Middle Atlantic Morocco, *Man* Northern Atlantic Morocco, *Op* Eastern High-Plateaux, *Om* Oriental mountains, *LM* Mediterranean coast, *R* Rif

Table 1 Phytosociological richness of the main vegetal formations

Vegetal formations	Associations or groupements	Alliances	Orders	Classes
Forests and preforests	142	14	5	2
Matorrals and spiny high-mountain xerophytes	97	16	7	3
Lawns, pozzines and meadows of high mountains	23	6	5	5
Rupicolous communities	42	11	5	3
Aquatic and hygrophile environments	57	4	5	3
Terrestrial halo-gypsophyle vegetation	114	10	6	5
Coastal, lagoon, estuarian or marine ecosystems	55	10	10	10
Saharan, steppic or pre-steppic vivace communities and Annual or perennial vegetation	140	26	23	13
Total	**670**	**97**	**66**	**44**

Contents

Forests and Preforests

Natural plant communities, dominated by trees and/or high shrubs, more or less dense. Hyper-humid, Humid, Sub-humid and Semi-arid bioclimates; warm, cool, temperate and cold winters. Inframediterranean to Supramediterranean belts.

Quercetea pubescentis Doing Kraft 1955 ex Scamoni and Passarge 1959

Oak, mixed deciduous and conifer (*Abies*, *Cedrus*) forests. Hyper-humid and Humid bioclimates. Supramediterranean belt.

Querco-Cedretalia atlanticae *Barbéro, Loisel & Quézel 1974*

Violo munbyanae-Cedrion atlanticae Barbéro, Quézel, and Rivas-Martinez 1981

- *Agropyro marginati-Cedretum atlanticae* Barbéro, Quézel, and Rivas-Martinez 1981

  Supramediterranean, Mountain mediterranean. Sub-humid; very cold winter. 1880–2000 m. Marly limestone.

  Eastern Middle-Atlas (Bou Iblane: Taffert) (Fig. 1).
- *Berberido hispanicae-Cedretum atlanticae* Benabid 1984

  Supramediterranean, Mountain mediterranean. Humid, Hyper-humid; cold to very cold winter. 1600–1900 m. Limestone.

  Western Rif (Central-eastern part of the “dorsale calcaire”).

  Subassociations:

  – *quercetosum rotundifoliae*
  – *geranietosum malviflorae*

M. S. Taleb, M. Fennane, *Vascular Plant Communities of Morocco*, Geobotany Studies, https://doi.org/10.1007/978-3-319-93704-5_1

Fig. 1 *Agropyro marginati-Cedretum atlanticae* (ph. Taleb)

- *Ficario ranunculoidis-Quercetum fagineae* Barbéro, Quézel, and Rivas-Martinez 1981
  Supramediterranean. Sub-humid, Humid; cold winter. 1400–1520 m. Shists, sandstone, quartz.
  Rif (jbel Tidighine).
- *Luzulo forsteri-Cedretum atlanticae* Barbéro, Quézel, and Rivas-Martinez 1981
  Supramediterranean, Mountain mediterranean. Humid, Hyper-humid. 1500–1900 m. Siliceous: schists. Deep forest brown soil.
  North Middle-Atlas (Tazekka) Rif (central Rif).
- *Paeonio maroccanae-Abietetum marocanae* Barbéro, Quézel, and Rivas-Martinez 1981
  Supramediterranean, Mountain mediterranean. Humid, Hyper-humid; cool, cold and very cold winter. 1500–1750 m. Limestone. Deep soils.
  Western Rif (region of Chaouen: jbel Tazzaot) (Fig. 2).
  Subassociations:

  – *abietetosum maroccanae*
  – *quercetosum rotundifoliae*

- *Polysticho setiferi-Prunetum lusitanicae* Barbéro, Quézel, and Rivas-Martinez 1981
  Supramediterranean. Humid; cold to very cold winter. 1480–1600 m. Limestone, siliceous.
  Ripisylva of the wadis and permanent sources of the Middle-Atlas (Tazekka) and the central western Rif.

Fig. 2 *Paeonio maroccanae-Abietetum marocanae*, Rif (massif Talassemtane) (ph. Ibn Tattou)

- *Trifolio ochroleuci-Quercetum fagineae* Benabid 1984

  Supramediterranean. Humid; cool and cold winter. 1400–1600 m. Limestone; earthy and deep soils.

  Western Rif (valleys and northern slopes of the "dorsale calcaire") (Fig. 3).

- *Violo munbyanae-Quercetum pyrenaicae* Benabid 1984

  Supramediterranean. Humid, Hyper-humid; cold winter. 1300–1700 m. Siliceous; earthy and deep.

  Summits of the centro-western Rif until jbel Outka toward the east and Tidighine toward the north.

  Many affinities with *Luzulo-Cedretum* wich appears in altitude toward the east.

  Subassociations:

  – *quercetosum pyrenaicae*

  – *quercetosum canariense*

Paeonio maroccanae-Cedrion atlanticae Barbéro, Quézel, and Rivas-Martinez 1981

- *Agyrocytiso battandieri-Cedretum atlanticae* Barbéro, Quézel, and Rivas-Martinez 1981

Fig. 3 *Trifolio ochroleuci-Quercetum fagineae*, Rif.

Supramediterranean, Mountain mediterranean. Humid; very cold winter. 1650–2000 m. Basalt, dolomite, limestone; deep soils.

Ravins and thalwegs of the tabular Middle-Atlas (Fig. 4).

- *Argyrocytiso battandieri-Quercetum rotundifoliae* Barbéro, Quézel, and Rivas-Martinez 1981

  Supramediterranean, Mountain mediterranean. Humid. 1900–2000 m. Cracked basalt.

  Central Middle-Atlas (south flank of the jbel Habri).

- *Paeonio maroccanae-Quercetum canariense* Barbéro, Quézel, and Rivas-Martinez 1981

  Supramediterranean, Mountain mediterranean. Humid; very cold winter. 1560–1730 m. Basalt, limestone, dolomite; deep forest brown soils, thick superficial humiferous horizon.

  Central Middle-Atlas (atlantic exposure, regions of El Hajeb, Azrou and Ifrane).

  Subassociations:

  - *quercetosum canariense*
  - *argyrocytisetosum battandieri*
  - *cedretosum atlanticae*

- *Piptathero paradoxi-Cedretum atlanticae* Quézel, Barbéro and Benabid 1987

  Supra- and Mountain mediterranean. Humid; very cold winter. 1800–2150 m. Marly-calcareous, rocky substrate.

Fig. 4 *Argyrocytiso battandieri-Cedretum atlanticae*, central Middle-Atlas (ph. Taleb)

Eastern High-Atlas (jbel Sloul, cliffs of the Maâsker, Mitkane, Cirque Jaâffar) (Fig. 5).

Quercetea ilicis Br.-Bl. ex A. Bolòs and O. de Bolòs in Bolos y Vayreda 1950

(*Pino halepensis-Quercetea ilicis* De Foucault and Julve 1991)

Evergreen forests and preforests of *Quercus, Tetraclinis, Juniperus, Argania, Pinus*... Humid, Sub-humid, Semi-arid and Arid bioclimates. Superior mediterranean, Mesomediterranean, Thermomediterranean and Inframediterranean belts.

Quercetalia ilicis *Br.-Bl. ex Molinier 1934 em. Rivas-Martinez 1975*

(*Quercetalia rotundifolio-ilicis* De Foucault and Julve 1991)

Fig. 5 *Piptathero paradoxi-Cedretum atlanticae*, Eastern High-Atlas (jbel Sloul) (ph. Taleb)

Oleo sylvestris-Quercion rotundifoliae Barbéro, Quézel, and Rivas-Martinez 1981

- *Myrto communis-Quercetum suberis* Barbéro, Quézel, and Rivas-Martinez 1981
  Thermomediterranean. Sub-humid, Humid; temperate to warm winter. 100–360 m. Sandstone.
  MA-1 (Bab Azhar) R-1-2 (Western Rif).
  Subassociations:
  - *myrtetosum communis*
  - *quercetosum cocciferae*
- *Tetraclino articulatae-Quercetum cocciferae* Benabid 1984 em. Benabid and Fennane 1994
  Thermomediterranean. Sub-humid. 130–360 m. Limestone, marls.
  North-western Rif.
- *Rusco hypophylli-Quercetum canariense* Rivas-Martinez 1974
  Thermomediterranean. Humid; warm winter. 80–350 m. Sandstone.
  Rif (Tangerian peninsula).
- *Rusco hypophylli-Quercetum cocciferae* Benabid 1984
  Thermomediterranean. Sub-humid, Humid. From sea level on atlantic slope, up to 900–1000 m in the Beni Snassen and Rif regions. Calcareous sandstone,

marly sandstone, argilaceous, schistous sandstone; soils brown forest, red ferralitic or rendzina.

Oriental mountains (Beni Snassen) Rif (Tangerian peninsula).

Nowadays represented only by small forest tasks in sacred sites.

Subassociations:

- *tetraclinetosum articulatae*
- *brachypodietosum sylvaticae*
- *myrtetosum communis*
- *luzuletosum forsteri*

• *Smilaci mauritanicae-Quercetum rotundifoliae* Barbéro, Quézel, and Rivas-Martinez 1981

Thermo- and Mesomediterranean. Sub-humid, Humid; temperate, hot winter. 400–900 m. Limestone, marl, sandstone, schist; superficial rocky soil.

Northern Morocco (region of Taza, southern reverse of the rifain ridges), Middle-Atlas and western High-Atlas.

• *Tamo communis-Oleetum sylvestris* Benabid 1984

Thermomediterranean. Semi-arid, Sub-humid, Humid; temperate to warm winter. 0–500 m. Marls, clays; vertisols.

Western Rif.

Oleaster formation, nowadays represented by samll tasks in sacred sites.

Subassociations:

- *ceratonietosum siliquae*
- *fraxinetosum angustifoliae*

• *Telino linifoliae-Quercetum suberis* Zeraia 1981

Thermomediterranean. Semi-arid, lower Sub-humid; temperate to hot winter. Sandy soils.

Northern atlantic Morocco (Maamora forest of *Quercus suber*) (Fig. 6).

• *Violo cochleatae-Fraxinetum angustifoliae* Benabid 1984

Thermomediterranean. Sub-humid, Humide; temperate winter. 400–700 m. Limestone-marl, dolomite-limestone, schist. Deep soil, rich, essentially colluvial and constantly moist without being hydromorphic.

Western Rif.

Balansaeo glaberrimae-Quercion rotundifoliae Barbéro, Quézel, and Rivas-Martinez 1981

• *Arbuto unedi-Quercetum rotundifoliae* Achhal 1986

Mesomediterranean. Sub-humid. 1000–1800 m. Granite, dolomite, basalt, schist; deep soil, rich, essentially colluvial and constantly moist without being hydromorphic.

Central High-Atlas (Fig. 7).

Subassociations:

Fig. 6 *Telino linifoliae-Quercetum suberis*, Maamora forest (ph. Taleb)

Fig. 7 *Arbuto unedi-Quercetum rotundifoliae* (ph. Taleb)

– *phillyretosum mediae*
– *quercetosum suberis*

- *Balansaeo glaberrimae-Cedretum atlanticae* Barbéro, Quézel, and Rivas-Martinez 1981

Supramediterranean, Mountain Mediterranean. Sub-humid, Humid; warm winter. 1600–2300 m. Limestone, basalt.

Central Middle-Atlas and Rif.
Mixed formations cedar/green oak.

- *Balansaeo glaberrimae-Quercetum canariense* Barbéro, Quézel, and Rivas-Martinez 1981
  Mesomediterranean, Supramediterranean. Humid. 1200–1500 m. Siliceous; deep colluvial soils.
  Northern Middle-Atlas (Tazekka).
  Subassociations:
  - *geranietosum malviflori*
  - *arbutetosum unedi*
- *Balansaeo glaberrimae-Quercetum rotundifoliae* Barbéro, Quézel, and Rivas-Martinez 1981
  Mesomediterranean, Supramediterranean. Sub-humid, Humid; temperate, cool winter. 750–1950 m.
  Middle-Atlas (Tazekka) and Rif.
  Subassociations:
  - *violetosum denhardtii*
  - *centaureetosum taganae*
  - *phlomidetosum samiae*
- *Cytiso triflori-Quercetum canariense* Benabid 1984
  Mesomediterranean. Humid, Hyper-humid; temperate winter. 1000–1400 m. Sandstone.
  Rif (west Rif, Outka).
  Subassociations:
  - *quercetosum canariense*
  - *quercetosum pyrenaicae*
- *Euphorbio briquetii-Quercetum rotundifoliae* Tregubov 1963
  Mesomediterranean. Sub-humid. 1000–1300 m. Schistous.
  Oriental mountains (high summits of Beni Snassen).
- *Festuco scaberrimae-Quercetum rotundifoliae* Quézel, Barbéro, Benabid, and Rivas-Martinez 1992a
  Mesomediterranean, Superior mediterranean. Sub-humid. 1000–1700 m. Compact limestone.
  Oriental mountains (mountains of Debdou, Bou Khouali and Hamza).
- *Genisto jahandiezii-Quercetum rotundifoliae* Barbéro, Quézel, and Rivas-Martinez 1981
  Meso- and Supramediterranean. Sub-humid; cold winter. 1400–1600 m. Limestone.
  Eastern Middle-Atlas (Taffert) (Fig. 8).
  Open forest with *Ampelodesma mauritanica* and *Genista quadriflora*; dense green oak formation with *Arabis nova* subsp. *iberica* and *Neottinea intacta*.
- *Luzulo atlanticae-Quercetum rotundifoliae* Barbéro, Quézel, and Rivas-Martinez 1981

Fig. 8 *Genisto jahandiezii-Quercetum rotundifoliae*, Eastern Middle-Atlas (ph. Taleb)

Meso- and Superior mediterranean. Semi-arid, Sub-humid, Humid. 1050–2050 m. Mainly siliceous: sandstone, schist, dolomite.

Central High-Atlas (regions of Aït Barka, Tizi-n-Test, Agaiouar, Taslida and Tizi-n-Zonguen).

Subassociations:

- *juniperetosum phoeniceae*
- *luzuletosum atlanticae*
- *cytisetosum balansae*

• *Paeonio maroccanae-Quercetum rotundifoliae* Barbéro, Quézel, and Rivas-Martinez 1981

Meso- and Supramediterranean. Humid, Hyper-humid. 1300–1600 m. Compact limestone, dolomitic limestone; brown red ferralitic soils.

Rif (Tazzaot mountain NW of Chaouen).

• *Sileno melliferae-Quercetum rotundifoliae* Quézel, Barbéro and Benabid 1987

Supra- and Mountain mediterranean. Sub-humid. 1500–2100 m. Limestone, marl, sandstone; rocky soils.

Eastern High-Atlas (mountains Sloul, Aberdouz, Hayim, Ayachi and Maasker).

• *Teucrio afrae-Quercetum suberis* Barbéro, Quézel, and Rivas-Martinez 1981

Meso- and Supramediterranean. Sub-humid, Humid. 1000–1500 m. Siliceous; brown soils, acide to moder.

Central Rif.

Mixed formations with oak suber abundant and green oak; degraded areas are marked by proliferation of *Cistus* spp.

Fig. 9 *Ampelodesmo mauritanicae-Chamaeropetum humilis* (ph. Taleb)

Pistacio lentisci-Rhamnetalia alaterni *Rivas-Martinez 1975*

(*Pino halepensis-Tetraclinetalia articulatae* De Foucault 1993)
(*Zizypho loti-Rhoetea oxyacanthae* De Foucault 1993 p.p.)

Asparago albidi - Rhamnion oleoidis Rivas Goday ex Rivas-Martinez (1975) and *Tetraclino articulatae-Pistacion atlanticae* Rivas-Martinez , Costa, and Itzco 1986

For these two syntaxa, the phytosociological value is not quite clear; their limits are matter of controverse depending on authors. Therefore, we prefer to present them together.

- *Ampelodesmo mauritanicae-Chamaeropetum humilis* Quézel, Barbéro, Benabid, and Rivas-Martinez 1992a

  Thermo- and Mesomediterranean. Sub-humid, Humid. 1100–1320 m. Limestone, marly calcareous.

  North-Eastern Middle-Atlas (Limestone plateau between Tazekka and the foothills of Bou-Iblane) (Fig. 9).

  Groupement belonging to the Mesomediterranean serial of green oak: degradation of *Genisto jahandiezii-Quercetum rotundifoliae*.

Fig. 10 *Bupleuro gibraltarici-Pinetum halepensis* (ph. Taleb)

- *Argyrolobio linnaeni-Pinetum halepensis* Achhal 1986
  Thermomediterranean. Semi-arid. 1700–2000 m. Sandstone, limestone.
  Central High-Atlas (High valley Agoundis).
- *Asparago aphylli-Calycotometum villosae* Rivas-Martinez 1975
  Thermomediterranean. Sub-humid, Humid. 30–150 m. Marly, siliceous.
  Northern Morocco (Tangerian peninsula).
  Matorral groupement, *Cistus monspeliensis* abundant.
- *Bupleuro gibraltarici-Pinetum halepensis* Tregubov 1963
  Thermomediterranean. Sub-humid. 650–850 m. Compact limestone; deep soils.
  Oriental mountains (Bni-Snassen: jbel Lakhdar) (Fig. 10).
- *Calycotomo intermediae-Oleetum sylvestris* Quézel, Barbéro, Benabid, Loisel, and Rivas-Martinez 1988
  Thermomediterranean. Arid, Semi-arid; temperate, warm winter. 80–320 m. Limestone, sandstone, marl; red soils.
  Littoral region of Al Hoceima and further towards the east.
  Preforest formation, proclimacic, still present only in small sacred areas.
- *Calycotomo intermediae-Tetraclinetum articulatae* Barbéro, Quézel, and Rivas-Martinez 1981
  Thermomediterranean. Sub-humid; temperate to warm winter. 400–1070 m. Limestone, marl.
  Northern Middle-Atlas (Tazekka) Rif (Tangerian peninsula between Sebta and Tangier).

Fig. 11 *Ceratonio siliquae-Tetraclinetum articulatae* (ph. Taleb)

Tetraclineria more or less degraded.
Subassociations:

- *tetraclinetosum articulatae*
- *pinetosum halepensis*

• *Ceratonio siliquae-Tetraclinetum articulatae* Fennane 1988
Thermomediterranean. Semi-arid, Sub-humid; temperate winter. 150–850 m. Limestone, marl.
Oriental mountains (Bni-Snassen, Debdou valley) (Fig. 11).
Proclimacic Tetraclineria.

• *Chamaeropo humilis-Rhamnetum lycioïdis* Bolos 1957
Thermomediterranean. Limestone.
Northern Morocco.

• *Clematidi cirrhosae-Ceratonietum siliquae* Barbéro, Quézel, and Rivas-Martinez 1981
Sub-humid; temperate, warm winter. 500–970 m. Limestone; rocky soil.
Western High-Atlas (jbel Amsitten), Middle-Atlas, North atlantic Morocco (central plateau), Rif.
Subassociations:

- *aristolochietosum baeticae*
- *ceratonietosum siliquae*

• *Coronillo valentinae-Pinetum halepensis* Quézel, Barbéro, Benabid, and Rivas-Martinez 1992a
Thermomediterranean. Sub-humid. 1000–1200 m. Sandstone, schist.

Fig. 12 *Cytiso fontanesii-Pinetum halepensis*, Eastern Middle-Atlas (ph. Taleb)

Oriental mountains (Debdou, Lalla Mimouna).

Mixed forest, dominated by two main species *Pinus halepensis* and *Quercus rotundifolia*.

- *Coronillo viminalis-Tetraclinetum articulatae* Barbéro, Quézel, and Rivas-Martinez 1981

  Thermomediterranean. Semi-arid. 420–670 m. Schist.

  Middle atlantic Morocco (Central plateau between Oued Zem and Oulmès).

  Subassociations:

  – *tetraclinetosum articulatae*

- *Cytiso fontanesii-Pinetum halepensis* Achhal 1986

  Thermomediterranean. Semi-arid, Sub-humid. 1400–1700 m. Calcareous marl.

  Eastern Middle-Atlas (Bou Nacer mountains), Central High-Atlas (between Amizmiz and Imi n'Tala).

  Edapho-climatic association, linked to marly calcareous substrates.

  Not reported by Benabid (2000) (Fig. 12).

- *Digitali laciniatae-Buxetum balearicae* Quézel, Barbéro, Benabid, Loisel, and Rivas-Martinez 1988

  Thermomediterranean. Sub-humid, Humid. 800–1100 m. Limestone.

  Northern and locally eastern slopes of the northern Rif chain, like on jbel Kelti, but also further south.

- *Genisto sparsiflorae-Tetraclinetum articulatae* Fennane 1988

  Thermomediterranean. Semi-arid.

  Middle atlantic Morocco (Haha, Ida-ou-Tanane).

Fig. 13 *Junipero phoeniceae-Tetraclinetum articulatae* (ph. Taleb)

Not reported by Benabid (2000).

- *Junipero oxycedri-Tetraclinetum articulatae* Fennane 1988

  Thermomediterranean. Semi-arid. 100–1300 m. Schist, marly calcareous, limestone, dolerites.

  North east of the Middle-Atlas (region of Bechchine south Taza) untill region of Jerada including north bordure of the oriental high plateaux, eastern Rif (south of Aïn Zorah, region of Anguied-Mezguittem).

- *Junipero phoeniceae-Pistacietum lentisci* Achhal 1986

  Thermomediterranean. Semi-arid;cold winter. 1350–1660 m. Limestone, sandstone, schist, basalt, quartzite.

  Central High-Atlas.

  Not reported by Benabid (2000).

- *Junipero phoeniceae-Tetraclinetum articulatae* Fennane 1982 em. 1988

  Inframediterranean. Semi-arid. 50–160 m. Calcareous sand, sandstone, calcareous sandstone.

  Middle atlantic Morocco (region of Essaouira, jbel Hadid NE of Essaouira) (Fig. 13).

- *Lavandulo dentatae-Tetraclinetum articulatae* Fennane 1988

  Thermomediterranean. Semi-arid; temperate winter. 900–1400 m. Limestone, calcareous marl, schist, dolerites, dolomitic limestone.

  High-Atlas (Nfiss valley, south slope of the High-Atlas).

Fig. 14 *Lavandulo dentatae-Tetraclinetum articulatae* (ph. Taleb)

Association showing many species of *Pistacio-Rhamnetalia*, but quite infiltraded by those of *Acacio-Arganietalia*, hence its originality (Fig. 14).

- *Lonicero implexae-Tetraclinetum articulatae* Fennane 1988

  Thermomediterranean. Semi-arid, Sub-humid. 480–930 m. Siliceous, pelites, limestone, sandstone, granit.

  Northern atlantic Morocco (Central plateau).

- *Periploco laevigatae-Tetraclinetum articulatae* Benabid 1984

  Thermomediterranean. Arid, Semi-arid. 30–350 m. Limestone, calcareous marl, schist.

  Rif (region Bou-Hmed, eastern Rif), Mediterranean coast (Littoral zone east of Al Hoceima) (Fig. 15).

  Proclimacic groupement.

  Subassociations:

  – *tetraclinetosum articulatae*
  – *arganietosum spinosae*

- *Phillyreo latifoliae-Oleetum sylvestris* Barbéro, Quézel, and Rivas-Martinez 1981

  Thermomediterranean. Semi-arid; warm winter. 300–700 m. Schist, sandstone, quatrz, flysch; deep soils.

  Middle atlantic Morocco (Central plateau: regions of khorifla, Khatouat, Sidi Bettache?).

- *Phillyreo latifoliae-Pistacietum lentisci* Benabid 1984

  Thermomediterranean. Sub-humid, Humid. 5–15 m. Sands.

Fig. 15 *Periploco laevigatae-Tetraclinetum articulatae*, Eastern Mediterranean coast (ph. Taleb)

Rif (region of Tangier, untill Moulay Bou-Selham).

Degraded groupement developed on the maritime dunes.

- *Phillyreo mediae-Tetraclinetum articulatae* Fennane 1988

  Thermomediterranean. Semi-arid; temperate winter. 100–750 m. Sandstone, pelite, schist, limestone.

  Northern atlantic Morocco (Central plateau: lower valleys of Beht, Bou Regreg and Grou in the countryside of Rabat-Casablanca).

- *Pistacio atlanticae-Ziziphetum loti* Benabid 1988

  Thermomediterranean. Semi-arid. 930–1000 m. Limestone, schist, sandstone.

  Northern atlantic Morocco (eastern Central plateau: plains and hills between Azrou and Khenifra) (Fig. 16).

- *Polygalo balansae-Tetraclinetum articulatae* Barbéro, Quézel, and Rivas-Martinez 1981

  Thermomediterranean. Semi-arid. 800–1400 m. Marls, limestones, schists, dolomites.

  Northern piedmont of the High-Atlas.

  Subassociations:

  – *tetraclinetosum articulatae*
  – *pinetosum halepensis*
  – *ceratonietosum siliquae*

- *Querco rotundifoliae-Tetraclinetum articulatae* Fennane 1988

  (*Arbuto unedi-Tetraclinetum articulatae* Fennane 1982)

Fig. 16 *Pistacio atlanticae-Ziziphetum loti* (ph. Taleb)

Thermomediterranean. Semi-arid, Sub-humid. 750–1350 m. Marl, sandstone, limestone, schist, calcareous marl, dolomite.

High-Atlas (from Ourika valley to the region of El Ksiba).

Association of upper horizons of Tetraclineria infiltrated with green oak and many mesophilic species.

Subassociations:

- *tetraclinetosum articulatae*
- *pinetosum halepensis*
- *arbutetosum unedi*
- *ceratonietosum siliquae*

• *Rhus pentaphyllae-Pistacietum atlanticae* Barbéro, Quézel, and Rivas-Martinez 1981

Thermomediterranean. Semi-arid; temperate winter. 300–730 m. Limestone, marl; superficial soils.

Northern atlantic Morocco (Central plateau).

Subassociations:

- *rhuscetosum pentaphyllae*
- *rhamnetosum oleoidis*

• *Rosmarino officinalis-Tetraclinetum articulatae* Fennane 1988

Fig. 17 *Tetraclino articulatae-Juniperetum turbinatae*, region of Berkine (ph. Taleb)

Thermomediterranean. Semi-arid; temperate winter. 800–1220 m. Limestone, marls, sandstone, pelites, marly calcareous.

Middle-Atlas (north and north east piedmont of Tazekka), Northern atlantic Morocco (Mdez valley).

Association with many species of *Rosmarinetalia.*

Subassociations:

- *tetraclinetosum articulatae*
- *juniperetosum phoeniceae*

• *Rosmarino tournefortii-Tetraclinetum articulatae* Fennane 1988

Thermomediterranean. Arid, Semi-arid. 220–1040 m. Marls, limestones, marly calcareous, sandstones, pelites.

Oriental mountains (Bni Snassen), Eastern plateaux (oriental high plateaux).

Association with many species of *Rosmarinetalia.*

Subassociations:

- *pinetosum halepensis*

• *Tetraclino articulatae-Juniperetum turbinatae* Quézel, Barbéro, Benabid, and Rivas-Martinez 1992a

Thermomediterranean. Semi-arid. 900–1100 m. Limestones, sandstones. Colluvial soils.

Eastern Middle-Atlas (Melloulou valley) (Fig. 17).

Subassociations:

- *pistacietosum lentisci*

Fig. 18 *Tetraclino articulatae-Euphorbietum resiniferae*, region of Bni-Mellal

– *oleetosum sylvestris*

• *Tetraclino articulatae-Euphorbietum resiniferae* Fennane 1988

Thermomediterranean. Aride, Semi-arid. 920–1050 m. Dolomite, limestone.

North piedmont of the High-Atlas (regions of Bezou, Tanant, Bin el-ouidane, Bni Mellal, El Ksiba) (Fig. 18).

Association marked by the abundance of *Euphorbia resinifera* beside *Tetraclinis articulata.*

• *Tetraclino articulatae-Jasminetum fruticantis* Fennane 1988

Thermomediterranean. Semi-arid; temperate winter. 600–1100 m. Dolomite, marls, limestone, sandstones.

Middle-Atlas (northern piedmont of Tazekka).

Association covering small areas, more or less isolated.

• *Tetraclino articulatae-Pinetum halepensis* Fennane 1988

Thermomediterranean. Semi-arid, Sub-humid. 820–1150 m. Calcareous schist, marly calcareous, pelites, schists, marls, sandstone.

Rif (region of Aknoul).

• *Helianthemo lavandulifoliae-Fumanetum calycinae* Peltier 1982

Thermomediterranean. 900–950 m. Gypsiferous marls, green marls.

Anti-Atlas (north flank of the jbel near Aït Chleuh), Western High-Atlas (Ida-ou-Tanane).

Association rich with species of *Rosmarinetalia.*

- *Lavandulo stoechae-Cistetum salviifoliae* Peltier 1982
  Thermomediterranean. 900–1100 m. Compact red clay, mixed with sandstone, doleritic basalts often vacuolar; superficial soil.
  Anti-Atlas (north flank of the jbel Ouaskal bordering the cuvette of Bigoudine), Western High-Atlas (Ida-ou-Tanane).
  Association rich with species of *Rosmarinetalia.*

Ericion arboreae Rivas-Martinez (1975) 1987

- *Cytiso arborei-Quercetum cocciferae* Quézel, Barbéro, Benabid, and Rivas-Martinez 1992a
  Thermomediterranean. Semi-arid. 750–1020 m. Limestone, dolomite, schist, granite.
  Oriental Mountains (Bni Snassen).
  Subassociations:

  – *ericetosum arborae*
  – *ampelodesmetosum mauritanicae*
  – *tetraclinetosum articulatae*

- *Erico arboreae-Myrtetum communis* Quézel, Barbéro, Benabid, Loisel, and Rivas-Martinez 1988
  Thermomediterranean. Sub-humid, Humid. 110–840 m. Schist, sandstone.
  Rif (regions of Tetouan, Oued Laou, Chaouen).
- *Erico arboreae-Quercetum cocciferae* Quézel, Barbéro, Benabid, and Rivas-Martinez 1992a
  Thermomediterranean. Semi-arid, Sub-humid. 550–1130 m. Sandstone, schist.
  Northern Middle-Atlas (Tazekka), Mediterranean coast (Karn mountain east of Al Hoceima).
  Preforest groupement: sparse suber oak; undergrowth with *Quercus coccifera, Cistus spp., Erica spp.*
  Subassociations:

  – *quercetosum cocciferae*
  – *quercetosum canariensis*
  – *eryngietosum tricuspidati*

- *Phillyreo latifoliae-Quercetum cocciferae* Quézel, Barbéro, Benabid, Loisel, and Rivas-Martinez 1988
  Thermomediterranean. Sub-humid, Humid. 400–1350 m. Sandstone.
  Rif (Rhomara and valley of Oued Nekor).
  Subassociations:

  – *phlomidetosum caballeroii*
  – *quercetosum rotundifoliae*

- *Phillyreo latifoliae-Arbutetum unedonis* Rivas-Goday and Galiano 1959
  Sub-humid, Humid.
  Northern Morocco.

For Benabid (2000: 261), this association belong to the alliances *Asparago-Rhamnion/Tetraclino-Pistacion*

Juniperion turbinatae Rivas-Martinez (1975) 1987

(*Juniperion lyciae* Barbéro, Quézel, and Rivas-Martinez 1981)

- *Clematido cirrhosae-Juniperetum lyciae* Barbéro, Quézel, and Rivas-Martinez 1981
  Thermomediterranean. Sub-humid. Sandy.
  Northern atlantic Morocco (maritime dunes of Mehdia).
- *Pino pinastri-Juniperetum lyciae* Barbéro, Quézel, and Rivas-Martinez 1981
  Thermomediterranean. Humid. 200–500 m. Sandstone, more or less consolidated sand.
  Rif (Punta Cires west of Tangier).
- *Rhamno rotundifolii-Juniperetum turbinatae* Quézel, Barbéro, Benabid, Loisel, and Rivas-Martinez 1988
  Thermomediterranean. Semi-arid. Sandy.
  North eastern Morocco (littoral zone of Saïdia).
  Juniperia, maritime dunes.

Junipero oxycedri-Rhamnion atlanticae Quézel and Barbéro 1986

- *Cisto laurifolii-Quercetum rotundifoliae* Achhal 1986
  Supramediterranean. Semi-arid. 2000–2400 m. Schist, sandstone.
  Central High-Atlas.
- *Coronillo ramosissimae-Quercetum rotundifoliae* Achhal 1986
  Mesomediterranean. Semi-arid. 1100–1800 m. Sandstone, limestone, schist.
  Central High-Atlas.
  Subassociations:

  – *sarothamnetosum arborei*
  – *pistacietosum terebenthii*

- *Festuco coeruleae-Quercetum rotundifoliae* Achhal 1986
  Mesomediterranean. Semi-arid. 1300–1800 m. Schist, sandstone, limestone.
  Central High-Atlas.
- *Junipero oxycedri-Pistacietum lentisci* Achhal 1986
  Meso- and Supramediterranean. Semi-arid; cold winter. 1350–1660 m. Schist, sandstone, basalt, limestone, quartzite; deep soils.
  Internal valleys of the Central High-Atlas.
- *Leuzeo coniferae-Pinetum halepensis* Quézel, Barbéro, and Benabid 1987
  Mesomediterranean. Semi-arid. 1250–1650 m. Marl, limestone.
  Eastern High-Atlas.
- *Cisto villosi-Telinetum segonnei* Peltier 1982

Fig. 19 *Cisto villosi-Telinetum segonnei*, Anti-Atlas (jbel Kest) (ph. Peltier)

Mesomediterranean. 1450–2000 m. Pleated sedimentary quartzites, silty, clayey silty, sandy silt.

Anti-Atlas (north flank of the jbel Kest, reliefs around Talmst, Asner, Ou-Gouguène, Imin-n-Taghzout, main summits of Tikwyène, Taskra, Amzlouye in the granitic massif of the Kerdous) (Fig. 19).

- *Cisto villosi-Genistetum ferocis* Peltier 1982

  Thermomeditarranean. 650–1450 m. Semi-arid. Quartzitic, rhyolitic, sandstone; deep soils.

  Western Anti-Atlas (West flank of the jbel Kest, reliefs around Aguert-ou-Dab, Amalou, Dar-Ourtane, Agoulmin, Oulbène, Tililout, Aougounz).

  Association of transition between *Pistacio-Rhamnetalia* and *Acacio-Arganietalia.*

Quercion fruticosae Barbéro, Quézel, and Rivas-Martinez 1981

- *Phillyreo angustifoliae-Quercetum fruticosae* Barbéro, Quézel, and Rivas-Martinez 1981

  Thermomediterranean. Humid. 1500–1800 m. Schistic.

  Rif (Tangerian Peninsula).

  Matorral: suberia (suber oak) and zeeneria (zeen oak) degraded.

Other Associations

- *Cytiso barbari-Ononidetum speciosae* Quézel, Barbéro, Benabid, Loisel, and Rivas-Martinez 1988

Mesomediterranean. Sub-humid. 1100–1500 m. Schists.
Central and eastern Rif.
Association little known, even by its authors (cf. Quézel et al. 1988: 83); not affiliated to any alliance.

Acacio gummiferae-Arganietalia spinosae *Barbéro, Benabid, Quézel, Rivas-Martinez, and Santos* 1982

(*Zizypho loti-Rhoetea oxyacanthae* De Foucault 1993 p.p.)

Senecio anteuphorbii-Arganion spinosae Barbéro, Benabid, Quézel, Rivas-Martinez, and Santos 1982

- *Artemisio huguetii-Traganopsidetum glomeratae* Barbéro, Benabid, Quézel, Rivas-Martinez, and Santos 1982
  Inframediterranean. 100–150 m. Arid; warm winter. Compact sandstone and sandstone boulders; stony soils.
  Littoral south of Sidi Ifni.
- *Crepido pinnatifidae-Warionietum saharae* Barbéro, Benabid, Quézel, Rivas-Martinez, and Santos 1982
  Inframediterranean. 280–300 m. Arid; warm winter. Quartzite or sandstone, rocky substrate.
  Littoral south of Sidi Ifni.
  Subrupicolous association.
- *Davallio canariensis-Dracaenetum ajgal* Benabid and Cuzin 1997
  Infra- and Thermomediterranean. 400–1400 m. Semi-arid, Sub-humid; temperate to warm winter. Quartzite. Cliffs.
  Anti-Atlas (borders of assif Amaghouz) (Fig. 20).
- *Digitario commutatae-Warionietum saharae* Barbéro, Benabid, Quézel, Rivas-Martinez, and Santos 1982
  Inframediterranean. 100–350 m. Arid; warm winter. Marly substrate; rocky soils with a high inclination.
  Middle atlantic Morocco and western High-Atlas (coastal region between Cape Tafelney and Oued Massa, extending inland to the N'Fiss valley).
- *Euphorbio beaumerianae-Arganietum spinosae* Barbéro, Benabid, Quézel, Rivas-Martinez, and Santos 1982
  Inframediterranean. 20–200 m. Semi-arid, Arid; warm winter. Generally limestone; superficial soils.
  Middle atlantic Morocco (littoral between Cape Tafelney and Oued Massa) (Fig. 21).
  Subassociations:

– *salsoletosum*
– *euphorbietosum regis-jubae*
– *genistetosum (ferox) microphyllae*

Fig. 20 *Davallio canariensis-Dracaenetum ajgal*, western Anti-Atlas (ph. Mangelsdorff in www.Teline.fr)

- *Euphorbio echini-Arganietum spinosae* Barbéro, Benabid, Quézel, Rivas-Martinez, and Santos 1982
  Inframediterranean. 120–400 m. Arid, Semi-arid; warm winter.
  Western Anti-Atlas, Middle atlantic Morocco (littoral region from Agadir to Sidi Ifni) (Fig. 22).
- *Helianthemo australis-Retametum monospermae* Benabid, Machrouh, and Schoenenberger 1994
  Infra- and Thermomediterranean. Semi-arid, Arid.
  Littoral hills between Safi and Sidi Ifni.
- *Limoniastro grandiflori-Arganietum spinosae* Quézel, Barbéro, Benabid, and Rivas-Martinez 1995
  250–750 m. Quatrzite.
  Inframediterranean (south limit), Thermomediterranean. Semi-arid, Arid. Colluvial spaces and thalweg in south of the western Anti-Atlas before oued Drâa (regions of Guelmim, Foum el-Hisn. . .).
  Subassociations:
  - *acacietosum raddianae*
  - *arganietosum spinosae*
- *Limonio mucronati-Traganetum moquinii* Benabid, Machrouh, and Schoenenberger 1994
  Inframediterranean. Semi-arid, Arid.
  Middle atlantic Morocco: association fixing the maritime dunes in the south of Agadir.

Fig. 21 *Euphorbio beaumerianae-Arganietum spinosae*, Anti-Atlas (ph. Taleb)

Fig. 22 *Euphorbio echini-Arganietum spinosae*, Anti-Atlas (ph. Taleb)

Fig. 23 *Periploco laevigatae-Arganietum spinosae*, western Anti-Atlas (ph. Taleb)

- *Penniseto dichotomae-Rhuscetum tripartitae* Barbéro, Benabid, Quézel, Rivas-Martinez, and Santos 1982
  Saharan. Sandy loam soils.
  Saharan Morocco (graras of Seguia el-Hamra; between Tarfaya and Laayoun).
- *Periploco laevigatae-Arganietum spinosae* Peltier and Msanda 1995
  Infra- and Thermomediterranean. 200–750 m. Arid, Semi-arid. Limestone, dolomite; docky soils.
  Western Anti-Atlas (Kerdous) (Fig. 23).
  Not reported by Benabid (2000).
  Subassociations:
  - *warionietosum saharae*
  - *convolvuletosum trabutiani*
  - *salsoletosum longifoliae*
  - *haloxylonetosum scopariae*
- *Periploco laevigatae-Juniperetum turbinatae* Benabid 2000
  Infra- or Thermomediterranean? Semi-arid, temperate winter.
  Middle atlantic Morocco (maritime dunes of Essaouira).
- *Polycnemo fontanesii-Arganietum spinosae* Peltier and Msanda 1995
  Thermomediterranean. Semi-arid. 700–900 m in north exposure, 800–1000 m in southern exposure. Limestone, dolomite.
  Anti-Atlas, replacing *Periploco-Arganietum* mid-altitude.

- *Traganopsidio glomeratae-Euphorbietum echini* Barbéro, Benabid, Quézel, Rivas-Martinez, and Santos 1982
  Saharan. Silty sand; rocky or relatively deep soils and talus slopes.
  Sahara (borders of the graras in Seguia al-Hamra).
  Subassociations:

  – *euphorbietosum echini*
  – *pulicarietosum lozanoi*

- Groupement with *Euphorbia balsamifera* var. *rogerii* Barbéro, Benabid, Quézel, Rivas-Martinez, and Santos 1982
  Saharan. Rocky sandy soils.
  Saharan Morocco (south margins of Seguia el-Hamra).

Acacion gummiferae Barbéro, Benabid, Quézel, Rivas-Martinez, and Santos 1982

- *Ephedro cossonii-Acacietum gummiferae* Barbéro, Quézel, and Rivas-Martinez 1981
  Thermomediterranean. Arid, Semi-arid. 700–1050 m. Limestone, marls, schists, dolomite.
  Middle atlantic Morocco (small sacred areas bordering the Haouz cuvette).
- *Ephedro cossonii-Arganietum spinosae* Barbéro, Benabid, Quézel, Rivas-Martinez, and Santos 1982
  Inframediterranean. Arid, Semi-arid; cold winter. 1000–1550 m. Acid rocks, often granite balls.
  Anti-Atlas.
- *Euphorbio echini-Tetraclinetum articulatae* Fennane 1988
  Inframediterranean. Arid, Semi-arid. 1050–1250 m. Compact limestone more or less fissured.
  Anti-Atlas (north of Bou Izakaren).
  The southernmost association of *Tetraclinis articulata.*
- *Hesperolaburno platyphylli-Arganietum spinosae* Barbéro, Benabid, Quézel, Rivas-Martinez, and Santos 1982
  Infra- and lower Thermomediterranean. Arid, Semi-arid; cool winter. 600–1100 m. Acid substrates.
  North western Anti-Atlas.
- *Oleo salicifoliae-Arganietum spinosae* Barbéro, Benabid, Quézel, Rivas-Martinez, and Santos 1982
  Infra- and lower Thermomediterranean. Semi-arid; temperate to warm winter. 400–1100 m. Limestone, sandstone, marls.
  High-Atlas (South-western and eastern flanks of the Ida-ou-Tanane massif) (Fig. 24).
- *Oleo salicifoliae-Tetraclinetum articulatae* Fennane 1988

Fig. 24 *Oleo salicifoliae-Arganietum spinosae*, western High-Atlas (Ida-ou-Tanane) (ph. Taleb)

Infra- and lower Thermomediterranean. Arid, Semi-arid. 250–1100 m. Limestone, dolomite, marly calcareous, marls, colluvions, sandstone.

High-Atlas (South-western and eastern flanks of the Ida-ou-Tanane massif).

- *Polygalo balansae-Arganietum spinosae* Barbéro, Benabid, Quézel, Rivas-Martinez, and Santos 1982

Infra- and lower Thermomediterranean. Arid, Semi-arid. 1000–1350 m. Limestone, marls; superficial and eroded soils.

Anti-Atlas, foothills of the High-Atlas in the north of Aoulouz.

Clear Argan forest, chamaephytes dominants.

- *Rubio longifoliae-Euphorbietum resiniferae* Barbéro, Quézel, and Rivas-Martinez 1981

Thermomediterranean. Semi-arid, temperate winter. 680–900 m. Limestone. Superficial soils.

High-Atlas of Demnate, Middle-Atlas (region of Beni Mellal).

- *Tetraclino articulatae-Arganietum spinosae* Fennane 1988

Infra- and Thermomediterranean. Arid, Semi-arid. 750–1300 m. Pelites, sandstones, limestones, marly-calcareous.

High-Atlas (Argana basin and southern slopes of the western High-Atlas) (Fig. 25).

Among the meridional tetraclinaies of Morocco, *Tetraclino-Arganietum* is distinguable by its heterogenous flora showing species of the two alliances *Acacion gummiferae* and *Ephedro-Juniperion*.

Fig. 25 *Tetraclino articulatae-Arganietum spinosae*, western High-Atlas (region of Argana) (ph. Taleb)

Other Associations

- *Aeonio arborei-Sonchetum pinnatifidi* Gehu and Biondi 1998
  Infra- or? Thermomediterranean. Semi-arid. Cliffs, rock walls cracked, steep slopes, very hard to access. warm exposures.
  Middle atlantic Morocco (north of Safi).
- *Arganio spinosae-Anagyrisetum foetidae* Peltier 1982
  Thermomediterranean. Semi-arid. 800–900 m. Limestone, marls. Deep soils, clayey.
  Middle atlantic Morocco (watershed of the Oued Souss: hills of Imouila).
  Argania is the dominant tree of the association.
- *Arganio spinosae-Artemisietum incultae* Peltier 1982
  Thermomediterranean. Arid. 800–1400 m. Limestone, dolomite. Shallow soils, colluvions.
  Anti-Atlas (Watershed of the Oued Souss: adrar Fouilim, adrar Minount, jbel Tanchkirra, jbel Taghzout).
- *Arganio spinosae-Cymbopogonetum schoenanthis* Peltier 1982
  Infra-, Thermomediterranean. 350 et 950 m. Arid, Semi-arid. Silty clay loam.
  High-Atlas (watershed of the Oued Souss: jbel Aouerir-Ichentane, jbel Bou-Addou, Aït Lahcen-ou-Saïd and other parts of the southern flank of the High-Atlas).

Fig. 26 *Arganio spinosae-Genistetum ferocis*, Anti-Atlas (ph. Taleb)

Clear formation with *Argania spinosa* and some rares *Tetraclinis articulata* and *Olea europaea* subsp. *oleaster.*

- *Arganio spinosae-Euphorbietum beaumierianae* Peltier 1982

  Inframediterranean. Semi-arid; warm winter. 50–300 m. Limestone, clay, marl, dolomite; rocky soils.

  Middle atlantic Morocco (Western and southern margins of the Ida-ou-Tanane massif from Agadir untill Cap Rhir).

  Subassociations:

  – *euphorbietosum rejis-jubae*
  – *rhuscetosum pentaphyllae*

- *Arganio spinosae-Euphorbietum echini* Peltier 1982

  Inframediterranean. Arid, Semi-arid. 350–1100 m. Schistous limestone, dolomite.

  Anti-Atlas (Aït Brahim, Aït Barka, Tinouainane, Tourhach, Tislane, Isk Timichcha, Abatoutou, etc.), Middle atlantic Morocco (lower Souss plain, along the road Tiznit-Agadir).

- *Arganio spinosae-Genistetum ferocis* Peltier 1982

  Infra- and lower Thermomediterranean. Semi-arid. 350–650 m. Limestone, dolomite; stoney on the surface, relatively deep soil.

  Anti-Atlas (northern fringes of the Kerdouss massif) (Fig. 26).

  Main arborescente/arbustive species: *Olea europaea* subsp. *oleaster*, *Ceratonia siliqua*, *Acacia gummifera.*

Fig. 27 *Arganio spinosae-Haloxyletum scopariae*, Anti-Atlas (ph. Taleb)

- *Arganio spinosae-Haloxyletum scopariae* Peltier 1982

  Infra- and lower Thermomediterranean. Arid, Semi-arid. 250–700 m. Deep loamy soils.

  Northern Anti-Atlas in the regions of Tazemmourt, Tinoainane, south of Taroudant (Bled El-Jdida, bled Tiouna, bled Bou Ghanem), etc. (Fig. 27).

  Mixed formation of *Argania spinosa* and *Acacia gummifera.*
- *Arganio spinosae-Hysperolaburnetum platycarpi* Peltier 1982

  Infra- and lower Thermomediterranean. Semi-arid. 350–1000 m. Limestone, dolomite. Ferralitic red soils.

  Anti-Atlas.
- *Arganio spinosae-Rhuscetum pentaphyllae* Peltier 1982

  Infra-, Thermomediterranean. Arid, Semi-arid. 600–1100 m. Schists, limestones, dolomites, yellow marls.

  Western High-Atlas (Ida-ou-Tanane, Tafingoult, Tagadirt n-ouchbarou, Tamaloukt, Taskat, adrar n-Tguilist, adrar n-Isseldaten Tagadirt . . .).

  Well represented in the Ida-ou-Tanane region.
- *Pistacio lentisci-Maytenetum senegalensae* Peltier 1982

  Thermomediterranean. 800–1250 m. Semi-arid. Rhyolites, granit, limestone, dolomite, schist. Deep soils of decarbonated loam.

  South flank of the Eastern High-Atlas (Surroundings of Aouzioua and jbel Bou-Taggount).

  Matorral more or less arborated.

Fig. 28 *Retamo dasycarpae-Juniperetum phoeniceae*, High-Atlas (ph. Taleb)

Ephedro majoris-Juniperetalia *Quézel and Barbéro (1981) 1986*

Ephedro majoris-Juniperion phoeniceae Quézel and Barbéro (1981) 1986

- *Coronillo ramosissimae-Juniperetum phoeniceae* Quézel and Barbéro 1981
  Thermomediterranean, Mesomediterranean. Arid, Semi-arid. 1000–1900 m. schist, Sandstone, dolomite.
  Central High-Atlas (N'Fiss valley).
- *Warionio saharae-Antirrhinetum ramosissimae* Quézel and Barbéro 1981
  Thermomediterranean. 1000–1300 m. Arid, Semi-arid. Rocky substrates.
  Western High-Atlas.
- *Retamo dasycarpae-Juniperetum phoeniceae* Quézel and Barbéro 1981
  Mesomediterranean, Superior mediterranean. Sub-humid; cold winter. 1300–2200 m. Marls, sandstone.
  Western High-Atlas (Fig. 28).
- *Zizipho loti-Rhuscetum tripartitae* Quézel, Barbéro, Benabid, and Rivas-Martinez 1994
  Thermomediterranean. Arid. 1100–1300 m. Limestone. Rocky substrates.
  Region of Errachidia, south flank of the Anti-Atlas: region of Tta-Tisgui ou Mallou.

Fig. 29 *Adenocarpo bacquei-Buxetum balearicae* subassociation retametosum, Eastern High-Atlas (ph. Taleb)

Junipero thuriferae-Quercion rotundifoliae Quézel and Barbéro (1981) 1986

- *Adenocarpo bacquei-Buxetum balearicae* Quézel and Barbéro 1981

  Thermo-, Meso- and Superior mediterranean. Arid, Semi-arid. 1600–2000 m. Limestone, marly-calcareous. Rocky substrate.

  Endemic of the Eastern High-Atlas (Dades cliffs; south flanks of the lower slopes of the jbel Aberdouz).

  Subassociations (Taleb and Fennane 2009):

  – *buxetosum balearicae*
  – *retametosum sphaerocarpae* (Fig. 29).

- *Berberido hispanicae-Fraxinetum dimorphae* Quézel and Barbéro 1981

  Superior and Mountain mediterranean. Semi-arid, Sub-humid; very cold winter. 1700–2200 m. Limestone, marl, stony or rocky substrate.

  Eastern High-Atlas (Ayachi, Maasker, National Parc of the Eastern High-Atlas), Central High-Atlas, Eastern Middle-Atlas (Fig. 30).

  Subassociations (Taleb and Fennane 2009):

  – *fraxinetosum dimorphae*
  – *juniperetosum thuriferae*

Fig. 30 *Berberido hispanicae-Fraxinetum dimorphae*, Eastern Middle-Atlas (ph. Taleb)

- *Bupleuro spinosi-Juniperetum phoeniceae* Quézel and Barbéro 1981

  Mountain mediterranean. Semi-arid; very cold winter. 1900–2800 m. Limestone, schist, quartzite. Rocky substrate.

  Anti-Atlas (jbel Saghro), Eastern High-Atlas (jbel Aberdouz), central High-Atlas.
- *Bupleuro spinosi-Juniperetum thuriferae* (Achhal 1986) Taleb and Fennane 2009

  (*Bupleuro spinosi-Juniperetum phoeniceae* Quézel and Barbéro 1981 *subassociation juniperetosum thuriferae* Achhal 1986)

  Mountain mediterranean, lower Oromediterranean. Semi-arid, Sub-humid; very cold winter. 2050–2600 m. Limestone, gabbro; rocky substrate.

  Eastern High-Atlas (jbel Aberdouz) (Fig. 31).

  Subassociation (Taleb and Fennane 2009):

  – *berberidetosum hispanicae*
- *Buxo balearicae-Quercetum rotundifoliae* Quézel and Barbéro 1981

  Superior and Mountain mediterranean. Semi-arid; cold to very cold winter. 1500–2200 m. Limestone, marls, rocky substrate.

  Eastern High-Atlas (Aberdouz, Ayachi and Masker mountains), Eastern Middle-Atlas (Bou-Nacer) (Fig. 32).

  Subassociations (Taleb and Fennane 2009):

Fig. 31 *Bupleuro spinosae-Juniperetum thuriferae*, Eastern High-Atlas (ph. Taleb)

Fig. 32 *Buxo balearicae-Quercetum rotundifoliae*, Eastern Middle-Atlas (Bou-Nacer) (ph. Taleb)

Fig. 33 *Crataego laciniatae-Berberidetum hispanicae*, Eastern High-Atlas (ph. Taleb)

- *buxetosum balearicae*
- *pinetosum halepensis*

- *Buxo sempervirentis-Juniperetum thuriferae* Quézel 1957
  Mountain mediterranean, Oromediterranean. Semi-arid; cold to very cold winter. 2200–2600 m. Sandstone, limestone. More or less fixed scree at the base of cliffs.
  High-Atlas (Akka n'Tazzert).
- *Crataego laciniatae-Berberidetum hispanicae* Quézel, Barbéro, Benabid, Loisel, and Rivas-Martinez 1992b
  Mountain mediterranean. Semi-arid, Sub-humid. 2000–2200 m. Limestone.
  Eastern High-Atlas (jbel Fazaz, Msadrid, Issoal), Eastern Middle-Atlas (Bou Nacer, Bou Iblane, Tichoukt) (Fig. 33).
- *Lonicero arboreae-Cedretum atlanticae* Barbéro, Quézel, and Rivas-Martinez 1981
  Mountain mediterranean, Oromediterranean. Sub-humid; cold and very cold winter. 2100–2350 m. Limestone, granit, dolomitic limestone; earthy soil, covered with rocks.
  Eastern High-Atlas (mountains Hayïm, Tazigzaout, Sloul, Taouarist, Ayachi and Masker), Eastern Middle-Atlas (Bou Nacer) (Fig. 34).
- *Ormenido scariosae-Quercetum rotundifoliae* Quézel and Barbéro 1981
  Supramediterranean, Mountain mediterranean. Semi-arid, Sub-humid; very cold winter. 1800–2450 m. Limestone, granit, marly-calcareous, rocky substrates.
  Oriental High-Atlas (valleys of Ahansal, Aït Bouguemez, Aït M'hamed, jbels Hayïm, Aberdouz, Ayachi, Maâsker) (Figs. 35 and 36).

Fig. 34 *Lonicero arboreae-Cedretum atlanticae*, Eastern Middle-Atlas (Bou-Nacer) (ph. Taleb)

Fig. 35 *Ormenido scariosae-Quercetum rotundifoliae*, Eastern High-Atlas (ph. Taleb)

Fig. 36 *Ormenis scariosa* (ph. Taleb)

- *Polygalo balansae-Quercetum rotundifoliae* Achhal 1986

  Mesomediterranean. Semi-arid, Sub-humid; cold winter. 1700–2000 m. Sandstone, schist.

  Central High-Atlas.
- *Retamo dasycarpae-Quercetum rotundifoliae* Achhal 1986

  Superior and Mountain mediterranean. 1600–2500 m, Semi-arid, Sub-humid; cold winter. Schist, dolomite, basalte, sandstone.

  South flank of the central High-Atlas.
- *Buxo balearicae-Juniperetum phoeniceae* Taleb and Fennane 2009

  Mesomediterranean, Superior mediterranean. Semi-arid; cold to cool winter. 1600–2000 m. Limestone; eroded soils.

  Eastern High-Atlas (Ziz valley, south flanks of the mountains Aberdouz, Ayachi and Masker) (Fig. 37).
- *Berberido hispanicae-Ribesetum uva-crispae* Taleb and Fennane 2009

  Mountain mediterranean. Semi-arid, Sub-humid; cold winter. 2050–2500 m. Earthy limestone.

  Eastern High-Atlas (mountains of Ayachi, Masker, Sloul, Tazigzaout, Taourist, Issoual, plateau des lacs), Eastern Middle-Atlas (Fig. 38).

Fig. 37 *Buxo balearicae-Juniperetum phoeniceae* (ph. Taleb)

Fig. 38 *Berberido hispanicae-Ribesetum uva-crispae*, Eastern Middle-Atlas (Bou-Nacer) (ph. Taleb)

Matorrals and Spiny High-Mountain Xerophytes

Calluno vulgaris-Ulicetea minoris Br.-Bl. and Tüxen ex Klika and Hadac 1944

Matorrals with *Erica spp.* Acidic soils, poor in nutrients. Hyper-humid, Humid and Sub-humid. Thermomediterranean.

Ulicetalia minoris *Quantin 1935*

Ericion umbellatae Br.-Bl., Silva, Rozeira, and Fontes 1952b em. Rivas-Martinez 1979

- *Erico scopariae-Stauracanthetum boivini* Quézel, Barbéro, Benabid, Loisel, and Rivas-Martinez 1988
  Thermomediterranean. Hyper-humid. 40–450 m. Sandstone, schistous marls. Rif (Tangerian peninsula).
  Degradation groupement of the *Phillyreo-Quercetum fruticosae*.
- *Erico umbellatae-Halimietum multiflori* Quézel, Barbéro, Benabid, Loisel, and Rivas-Martinez 1988
  Thermomediterranean. Sub-humid; temperate to warm winter. 150–350 m. Sandy soil, rich in silica.
  Northern atlantic Morocco (Gharb: regions of Larache and Arbaoua).
  Degradation groupement of the *Myrto communis-Quercetum suberis*.
- *Stauracantho boivini-Drosophylletum lusitanicae* Quézel, Barbéro, Benabid, Loisel, and Rivas-Martinez 1988
  Thermomediterranean. Humid, Hyper-humid. 100–200 m. Sandstone, rocky. Rif (Tangerian peninsula).
  Degradation groupement of the *Erico scopariae-Stauracanthetum boivini*.

M. S. Taleb, M. Fennane, *Vascular Plant Communities of Morocco*, Geobotany Studies, https://doi.org/10.1007/978-3-319-93704-5_2

Fig. 1 Matorral of *Cisto-Lavanduletea*, siliceous substrate, central Rif (ph. Ibn Tattou)

***Genistion micrantho-anglicae* Rivas-Martinez 1979**

- *Genisto anglicae-Ericetum ciliaris* Quézel, Barbéro, Benabid, Loisel, and Rivas-Martinez 1988
  Thermomediterranean. Humid, Hyper-humid. 100–400 m.
  Rif (borders of peatlands in the area between Larache, Chaouen and Tangier).

Cisto-Lavanduletea stoechadis Br.-Bl. in Braun-Blanquet, Molinier and Wagner 1940

Matorrals (maquis) resulting from the degradation of tree communities of *Quercetea ilicis and Quercetea pubescentis* (Fig. 1). Siliceous, non calcareous substrates. Hyper-humid, Humid, Sub-humid. Supramediterranean to Thermo-mediterranean.

Halimietalia riphaeo-atlantici *Quézel, Barbéro, Benabid, Loisel, and Rivas-Martinez 1988*

***Erico riphaeae-Cistion varii* Quézel, Barbéro, Benabid, Loisel, and Rivas-Martinez 1988**

- *Cistetum populifolio-varii* Quézel, Barbéro, Benabid, Loisel, and Rivas-Martinez 1988
  Meso- and Supramediterranean. Humid. 1200–1500 m. Sandstone.

Rif (jbel Sougna, jbel Bou Hachem).
Degradation of deciduous formation of *Cytiso triflori-Quercetum canariense.*
Subassociations:

- *halimietosum viscosi*
- *quercetosum pyrenaicae*

- *Ericetum riphaeo-umbellatae* Quézel, Barbéro, Benabid, Loisel, and Rivas-Martinez 1988
  Supramediterranean. Hyper-humid. 1000–1600 m. Sandstone.
  Rif (high plateau of Bou Hachem).
  Degradation groupement of the *Violo munbyanae-Quercetum pyrenaicae.*
- *Halimio angustifolii-Cistetum crispi* Quézel, Barbéro, Benabid, Loisel, and Rivas-Martinez 1988
  Thermomediterranean. Sub-humid, Humid. 500–1100 m. Sandstone.
  Rif (Inner basins of the western Rif, region of Chaouen).
  Advanced stage of degradation of the *Myrto communis-Quercetum suberis.*

Halimio-Cistion atlanticae Deil 1984

- *Cisto atlantici-Genistetum quadriflorae* Deil 1984 em. Quézel et al. 1988
  (*Halimio atlantici-Genistetum quadriflorae* Benabid 1982)
  Supra- and Mountain mediterranean. Sub-humid, Humid. 1300–2100 m. Schists, quartz.
  Northern Middle-Atlas (jbel Tazekka), Rif (jbel Tidighine until Tizi Ifri).
  Degradation stage of cedar forest.
  Subassociations:

- *thymetosum riatari*
- *ericetosum arborae*
- *cistetosum ladaniferi*
- *cistetosum albidi*
- *scorzoneretosum pygmae*

- *Genistello tridentatae-Teucrietum oxylepidi* Deil 1984
  Supramediterranean, Mountain mediterranean. Sub-humid, Humid; cold and cool winter. 1600–2450 m. Siliceous.
  Northern Middle-Atlas (jbel Tazekka), Rif (central siliceous part, jbel Tidighine, jbel Erz).
  Subassociations:

- *telephietosum imperati*
- *thymetosum riatari*

- *Genisto eriocladae-Cistetum ladaniferi* Quézel, Barbéro, Benabid, Loisel, and Rivas-Martinez 1988
  Meso- and Supramediterranean. Sub-humid. 1300–1500 m. Siliceous.
  Eastern Rif (between Targuist and northern Aknoul).

- *Halimio atlantici-Stauracanthetum boivinii* Deil 1984
  Mesomediterranean. 1000–1700 m. Eroded schists.
  Northern Middle-Atlas, central Rif.
  Degradation groupement of *Teucrio-Quercetum suberis*, of *Juniperus oxycedrus* and of *Quercus rotundifolia.*
  Association not accepted by Quézel et al. (1988: 89).
- *Halimio viscosi-Lavanduletum atlanticae* Quézel, Barbéro, Benabid, Loisel, and Rivas-Martinez 1988
  Mesomediterranean, Supramediterranean. Sub-humid. 1300–1500 m. Shists; superficial soils.
  Rif (inner zones of the central siliceous part, high valley of oued Sra, particularly near Tleta Ketama).
- *Stauracantho tazzensis-Ericetum riphaeae* Quézel, Barbéro, Benabid, Loisel, and Rivas-Martinez 1988
  Meso- and Supramediterranean. Humid, Hyper-humid. Siliceous; deep soils.
  Rif (summits and deforested areas between Bab Berred and Ketama).
  Degradation groupement of the *Teucrio afrae-Quercetum suberis* and *Cytiso triflori-Quercetum canariense.*
  Subassociations:

  – *cistetosum atlantici*
  – *festucetosum durandoi*

- *Cytisetum megalanthi* Deil 1984
  Supramediterranean. 1200–1800 m. Siliceous; deep soils.
  Rif (from jbel Tizirène and jbel Outka until Tizi Ifri and jbel Tamchacht).
  Association not accepted by Quézel et al. (1988: 89).

Lavanduletalia stoechadis *Br.-Bl. in Braun-Blanquet et al. 1940*

Genisto quadriflorae-Lavandulion atlanticae Benabid 1988

- *Genisto quadriflorae-Lavanduletum atlanticae* Benabid 1988
  Mesomediterranean. Sub-humid. 1250–1350 m. Siliceous; eroded soils.
  Middle-Atlas/ Northern atlantic Morocco (region of Ain-Leuh, transition zone between the Middle-Atlas and the Central Plateau).
- *Halimio villosissimae-Genistetum quadriflorae* Quézel, Barbéro, Benabid, Loisel, and Rivas-Martinez 1992b
  Mesomediterranean. Sub-humid. Quartzitic rocks.
  Northern atlantic Morocco (regions of Oulmes/El-Harcha/Moulay Bou-Azza).
  Degradation stage of the mixed serial: green oak - cork oak.
  Citation after Benabid (2000: 267); doubtful informations, must be confirmed, cf. Quézel, Barbéro, Benabid, Loisel, and Rivas-Martinez 1988 (p. 89).

Ononido-Rosmarinetea Br.-Bl. in O. Bolos y Vayreda 1950

Secondary communities resulting from the degradation of the forests and preforests formations of the *Quercetea ilicis*. Base-rich substrates, calcareous, dolomitic... Humid, Sub-humid, Semi-arid, Arid. Superior mediterranean to Thermomediterranean.

Rosmarinetalia officinalis *Br.-Bl. ex Molinier 1934*

Lithodoro maroccanae-Ulicion funkii Quézel, Barbéro, Benabid, Loisel, and Rivas-Martinez 1988

- *Cisto crispi-Sanguisorbetum mauritanicae* Quézel, Barbéro, Benabid, Loisel, and Rivas-Martinez 1988
  Thermomediterranean. Sub-humid, Humid. 300–550 m. Limestone.
  Rif (region of Tetouan-Sebta: jbel Moussa, Dar Ben Karrich).
  Quercus coccifera serial; potential association: *Rusco hypophylli-Quercetum cocciferae*.
- *Phlomido caballeroi-Micromerietum inodorae* Quézel, Barbéro, Benabid, Loisel, and Rivas-Martinez 1988
  Thermomediterranean. Semi-arid, Sub-humid. 650–1450 m.
  Rif (lower valley of Oued Laou).
- *Pino maghrebianae-Ulicetum funkii* Quézel, Barbéro, Benabid, Loisel, and Rivas-Martinez 1988
  Thermo- and Mesomediterranean. Sub-humid, Humid. Dolomite, sandy substrate.
  Rif (jbel Kelti, jbel Lakraâ).
  Degradation facies of the edapho-climacic association *Pino maghrebianae-Quercetum rotundifoliae*.
- *Saturejo graecae-Coridothymetum capitati* Quézel, Barbéro, Benabid, Loisel, and Rivas-Martinez 1988
  Thermomediterranean. Sub-humid, Humid. 150–350 m. Limestone; eroded soils.
  Rif (region of Tetouan-Sebta: jbel Moussa, Dar Ben Karrich).
  Ultimate stage of degradation of the *Quercus coccifera* serial; potential association: *Rusco hypophylli-Quercetum cocciferae*.

Pseudoscabioso grosii-Origanion grosii Quézel, Barbéro, Benabid, Loisel, and Rivas-Martinez 1988

- *Anthyllido polycephalae-Stachydetum fontqueri* Quézel, Barbéro, Benabid, Loisel, and Rivas-Martinez 1988
  Meso- and Superior mediterranean. Sub-humid, Humid. 950–1500 m. Compact dolomite.
  Rif (Talassemtane, jbel Kelti).

Fig. 2 *Anarrhino fruticosi-Globularietum nainii*, Eastern High-Atlas (ph. Taleb)

- *Pino maghrebianae-Genistetum gomaricae* Quézel, Barbéro, Benabid, Loisel, and Rivas-Martinez 1988

  Supramediterranean. Humid, Hyper-humid. 1600–1700 m. Dolomite.

  Rif.

  Pedolimax. Supramediterranean serial of *Abies maroccana*.
- *Poo ligulatae-Ononidetum jahandiezii* Quézel, Barbéro, Benabid, Loisel, and Rivas-Martinez 1992b

  Supramediterranean, Mountain mediterranean. Hyper-humid. 1600–1850 m. Dolomite.

  Rif (massif of Talassemtane-Madissouka).

  Lawns, dominated by woody chamephytes.

Anarrhino fruticosi-Astragaletalia armati *Quézel, Barbéro, Benabid, Loisel, and Rivas-Martinez 1992b*

Bupleuro aiouensis-Globularion nainii Quézel, Barbéro, Benabid, Loisel, and Rivas-Martinez 1992b

- *Anarrhino fruticosi-Globularietum nainii* Quézel, Barbéro, Benabid, Loisel, and Rivas-Martinez 1992b

  Mesomediterranean, Superior mediterranean. Semi-arid. 1540–2250 m. Limestone, marls, sandstone, rocky substrates.

  Eastern High-Atlas (Fig. 2).

  Low floristic richness.

Fig. 3 *Buffonio murbeckii-Lotononidetum tapetiformis*, Eastern High-Atlas (ph. Taleb)

- *Arenario parviflorae-Sixalidetum parielii* Quézel, Barbéro, Benabid, Loisel, and Rivas-Martinez 1992b

  Mountain mediterranean. Semi-arid; cold winter. 2000–2200 m. Limestone, marls, rocky substrates.

  Eastern High-Atlas (south flank of the Jbel Maâsker).

  Edapho-climacic association.

- *Artemisio mesatlanticae-Genistetum pomariensis* Quézel, Barbéro, Benabid, Loisel, and Rivas-Martinez 1992b

  Mesomediterranean, Superior mediterranean. Semi-arid; cold winter. 1400–1900 m. Limestone, marly limestone, colluvions, dolomite; deep soils.

  Eastern Middle-Atlas (between Boulemane and Imouzzer Marmoucha, and locally towards the east on the south piedmonts of Bou Nacer).

  Presteppic serial of the green oak.

  Subassociations:

  – *artemisietosum mesatlanticae*
  – *buxetosum balearicae*

- *Buffonio murbeckii-Lotononidetum tapetiformis* Quézel, Barbéro, Benabid, Loisel, and Rivas-Martinez 1992b

  Mountain mediterranean. Semi-arid, Sub-humid. 1600–2300 m. Marls, marly-calcareous, schistous.

  Eastern High-Atlas (Anemzi, Zaouiat Sidi Hamza, between Tounfite and Tagoudite, between Agoudim and Anemzi, north and south lower slopes of jbel Aberdouz, south slopes of jbel Ayachi) (Figs. 3 and 4).

  Berberido hispanicae-Fraxinetum dimorphae serial.

Fig. 4 *Lotononis tapetiformis* (ph. Taleb)

- *Lavandulo brevidentis-Hertietum maroccanae* Quézel, Barbéro, Benabid, and Rivas-Martinez 1994

  Eastern High-Atlas (difficult limestone escarpments).
- *Ormenido africanae-Centaureetum benoistii* Quézel, Barbéro, Benabid, Loisel, and Rivas-Martinez 1992b

  Superior mediterranean. Semi-arid; cold to very cold winter. 1700–2100 m. Compact limestones, marls.

  Eastern High-Atlas (*probably endemic of the south-eastern region of Midelt*).

  Subassociations:

  – *rosmarinetosum officinalis*
  – *jurinetosum humilis*

- *Salvio mesatlanticae-Teucrietum mideltensis* Quézel, Barbéro, Benabid, Loisel, and Rivas-Martinez 1992b

  Mesomediterranean, Superior mediterranean.1600–1800 m. Semi-arid; cold to very cold winter. Sandstone, pelites.

  Eastern High-Atlas (northern flank of the jbel Ayachi).
- *Scabioso condensatae-Astragaletum exulis* Quézel, Barbéro, Benabid, Loisel, and Rivas-Martinez 1992b

  Superior mediterranean. 1800–1900 m. Semi-arid; cold to very cold winter. Marly-calcareous.

  Eastern High-Atlas (north flank of the jbel Maâsker in Tounfite region), Eastern Middle-Atlas (eastern border of the jbel Tichoukt, particularly in Aïn Makhlouf region).

  Association known only in small areas.

- *Siderito jahandiezii-Erinaceetum anthyllidis* Quézel, Barbéro, Benabid, Loisel, and Rivas-Martinez 1992b

    Mesomediterranean and Superior mediterranean? 1500–1800 m. Limestone.
    Windy exposures in the High Moulouya.
    Association little known (Quézel, Barbéro, Benabid, Loisel, and Rivas-Martinez 1992b: 144).

- *Teucrio ducellieri-Artemisietum incultae* Quézel, Barbéro, Benabid, Loisel, and Rivas-Martinez 1992b

    Mesomediterranean. 1450–1550 m. Arid; cold winter. Limestone scree; stony, rocky soils.
    Eastern plateaux (high Moulouya in the region of Midelt-Itzer).
    Like steppic formation, rather poor in species. The potential vegetation seems correspond to a red juniperia.

Lavandulo mairei-Carthamion fruticosi Quézel, Barbéro, Benabid, and Rivas-Martinez 1994

- *Adenocarpo leiocarpi-Retametum dasycarpae* Quézel, Barbéro, Benabid, and Rivas-Martinez 1994

    Meso- and Supramediterranean. Arid, Semi-arid. 1400–1900 m. Marls.
    High-Atlas (south flank of the Mgoun chain).
    Thermophile and xerophile.

- *Convolvulo trabutiani-Genistetum capitellatae* Quézel, Barbéro, Benabid, and Rivas-Martinez 1994

    Mesomediterranean. 1100–1300 m. Semi-arid; cold winter. Limestone.
    Eastern High-Atlas (High-Atlas piedmonts in the region of Errachidia-Gourrama).
    Subassociations:

    - *teucrietosum malenconiani*
    - *genistetosum capitellatae*

- *Genisto myrianthae-Carthametum fruticosi* Quézel, Barbéro, Benabid, and Rivas-Martinez 1994

    Meso- and Superior mediterranean. 1300–1800 m. Arid, Semi-arid; cold winter. Limestone.
    Association covering a large area approximately limited by Rheriss valley in the east, Tazert in the west, High-Atlas in the north and the Saghro in the south.
    Subassociations:

    - *globularietosum arabicae*
    - *adenocarpetosum bacquei*
    - *anvilletosum radiatae*
    - *astragaletosum armatae*

- *Thymo commutati-Rosmarinetum officinalis* Quézel, Barbéro, Benabid, and Rivas-Martinez 1994
  Thermo- and lower Mesomediterranean. 1200–1750 m. Arid; cold winter. Limestone, marly-calcareous; rocky soils.
  Eastern High-Atlas (medium part of the Ziz valley).

Other Associations

- *Astragalo armati-Bupleuretum atlanticae* Nègre 1961
  Mesomediterranean, Superior mediterranean. Sub-humid; cold to very cold winter. Limestone more or less sandy, marly limestone.
  Central Middle-Atlas (jbel Saa).
  Subassociations:

  – *carduncellusetosum pomeliani*
  – *leontodonetosum pitardii*

- *Perralderio paui-Halogetonetum alopecuroidis* Deil 2005
  Thermomediterranean.
  Northern Morocco: regions of Taza and Al Hoceima (Tazekka, Tizi Ouzli, Msoun, Col of Touahar, Aknoul, Midar, Tsoul...).

Cisto mauritanici-Thymetalia munbyani *Quézel, Barbéro, Benabid, Loisel, and Rivas-Martinez 1992b*

Ulici africanae-Rosmarinion tournefortii Quézel, Barbéro, Benabid, Loisel, and Rivas-Martinez 1992b

- *Artemisio incultae-Rosmarinetum tournefortii* Quézel, Barbéro, Benabid, Loisel, and Rivas-Martinez 1992b
  Thermomediterranean. 200–700 m. Arid, Semi-arid; cold winter. Limestone, sandstone.
  Eastern plateaux (Taourirt region on the reliefs bordering the low valley of Oued Za); certainly endemic.
- *Artemisio incultae-Sideritetum briquetianae* Quézel, Barbéro, Benabid, Loisel, and Rivas-Martinez 1992b
  Thermo- and lower Mesomediterranean. 400–700 m. Arid, Semi-arid; cool winter. Sandstone, calcareous sandstone, gypseous marls.
  Eastern Morocco (Saka region on the Moulouyan valley border).
- *Genisto moulleronii-Ericetum multiflorae* Quézel, Barbéro, Benabid, Loisel, and Rivas-Martinez 1992b
  Thermomediterranean. Semi-arid. 600–800 m. Marly limestone.
  Oriental mountains (western Beni Snassen, abundant in jbel Lakhdar).
  Degradation groupement of the *Bupleuro gibraltarici-Pinetum halepensis*.

- *Genistetum ramosissimo-demnatensis* Achhal, Ziri and Khattabi 2004
  Thermomediterranean. 850–1100 m. Sub-humid; temperate winter. Doleritic, limestone, clayey.
  Oriental mountains (eastern Bni Snassen between Tinissene and Jbel Farouane).
- *Helianthemo caput-felis-Cistetum heterophylli* Quézel, Barbéro, Benabid, Loisel, and Rivas-Martinez 1988
  Thermomediterranean. 200–450 m. Semi-arid; temperate to warm winter. Volcanic substrate.
  Mediterranean coast (regions of Melilla–Nador).
  Degradation stage of the *Periploco-Tetraclinetum*, showing some *Pinus halepensis* peuplements.
  Subassociations:
  - *genistetosum quadriflorae*
  - *Helianthemetosum origanifoliae*
- *Lavandulo dentatae-Genistetum durieui* Quézel, Barbéro, Benabid, Loisel, and Rivas-Martinez 1992b
  (=? *Lavandulo-Callitrietum* Tregubov 1963)
  Thermomediterranean, locally Mesomediterranean. 500–900 m. Sub-humid; temperate winter. Limestone, marls.
  Oriental mountains (Bni-Snassen). Certainly present also in western Algeria (region of Oran).
  Degradation groupement of Thermomediterranean serial of *Ceratonia-Tetraclinis*.
- *Salvio aegyptiacae-Thymetum glandulosi* Quézel, Barbéro, Benabid, Loisel, and Rivas-Martinez 1988
  Thermomediterranean. 80–200 m. Semi-arid; warm winter. Limestone, sandstone.
  Mediterranean coast (massif of Bokkoya).
  Degradation stage of the *Periploco-Tetraclinetum*.
- *Ulici africanae-Cistetum sericei* Quézel, Barbéro, Benabid, Loisel, and Rivas-Martinez 1992b
  Upper Thermomediterranean. 800–900 m. Sub-humid; temperate to cool winter. Sandstone, limestone.
  Oriental mountains (Bni Snassen). Maybe present also in Algeria (region of Oran)

Degradation groupement of the *Ceratonio-Tetraclinetum* and/or? the *Pistacio lentisci-Quercetum cocciferae*.

***Staehelino macrocephalae-Genistion pseudoretamoidis* Quézel, Barbéro, Benabid, Loisel, and Rivas-Martinez 1992b**

- *Coronillo junceae-Genistetum atlanticae* Quézel, Barbéro, Benabid, Loisel, and Rivas-Martinez 1992b

  Upper Thermomediterranean. 700–1000 m. Semi-arid; temperate to cool winter. Limestone, and locally acid rocks.

  Eastern plateaux (piedmonts of Gada Debdou). Maybe present also in Algeria (mountains of Telemcen).
- *Genistetum eriocladae-pseudoretamoidis* Quézel, Barbéro, Benabid, Loisel, and Rivas-Martinez 1992b

  Thermomediterranean. 900–1100 m. Semi-arid; cool winter. Limestone, marl.

  Estern Middle-Atlas (Bou Iblane in the medium valley of Melloulou).
- *Genistetum intermedio-quadriflorae* Quézel, Barbéro, Benabid, Loisel, and Rivas-Martinez 1992b

  Upper Thermomediterranean, lower Mesomediterranean. 1100–1500 m. Sub-humid; cool to cold winter. Sandstony marls.

  Eastern plateaux (high valley of the Moulouya), Eastern Middle-Atlas (north flank of the Bou Iblane).

  Association showing some peuplements of *Pinus halepensis* and *Pinus maghrebiana*.
- *Hedysaro subspeciosi-Thymetum maroccani* Quézel, Barbéro, Benabid, Loisel, and Rivas-Martinez 1992b

  Mesomediterranean. 1120–1500 m. Sub-humid; cool to cold winter. Limestone.

  Middle-Atlas / Eastern plateaux (south flank of the Bou Iblane).

  Degradation stage of the *Genisto jahandiezii-Quercetum rotundifoliae.*
- *Pino halepensis-Genistetum pseudoretamoidis* Quézel, Barbéro, Benabid, Loisel, and Rivas-Martinez 1988

  Thermomediterranean. 800–950 m. Semi-arid. Marl and flyschs; eroded soils.

  Rif (region of Aknoul).

  Degraded formation, *Pinus halepensis* and *Tetraclinis articulata* are in general present.
- *Ptilostemo leptophylli-Fontquerietum pauii* Quézel, Barbéro, Benabid, Loisel, and Rivas-Martinez 1988

  Mesomediterranean. 850–1150 m. Semi-arid; temperate winter. Marls.

  Eastern Rif / Mediterranean coast (medium and high Nekkor, south-eastern rifian hills).

  Closely linked to marl substrates.

  This association should be abandoned, "the name-giving species have quite different ecological requirements" (Deil 2005)
- *Stipo lagascae-Linetum lycopodioidis* Quézel, Barbéro, Benabid, Loisel, and Rivas-Martinez 1988

  Mesomediterranean. 1500–1600 m. Sub-humid; cool winter. Limestone.

  Eastern Rif (northern Aknoul in the Azrou Akechar region).

  Sub-rupicolous association, closely linked to limestone substrates. Belong to the serial of green oak.

Erinacetalia anthyllidis *Quézel 1952*

Diantho maroccani-Astragalion maroccani Quézel, Barbéro, Benabid, Loisel, and Rivas-Martinez 1988

(*Berberidion hispanicae* Benabid 1984).

- *Astragaletum numidico-maroccani* Quézel, Barbéro, Benabid, Loisel, and Rivas-Martinez 1988
  Mountain mediterranean. 1700–2000 m. Humid, Hyper-humid; cold winter. Limestone; medium and coarse scree.
  Western Rif (north and eastern flanks of jbel Lakraâ and jbel Kelti).
- *Ononido aragonensis-Ptilotrichetum spinosi* Quézel, Barbéro, Benabid, Loisel, and Rivas-Martinez 1988
  Mountain mediterranean. 1900–2150 m. Humid, Hyper-humid; cold winter. Compact limestone.
  Western Rif (jbel Lakraâ and jbel Kelti).

Genisto pseudopilosae-Thymion comosi Benabid 1988

- *Arenario armerinae-Adenocarpetum boudyi* Benabid 1988
  Superior and Mountain mediterranean. 1700–1850 m. Humid; cold winter. Sandy-dolomite; rocky soil.
  Central Middle-Atlas (Around Aïn Leuh).
- *Armerio choulettianae-Genistetum pseudopilosae* Nègre 1961
  (*Dasypiro hordeacei-Genistetum pseudopilosae* Benabid 1988)
  Superior and Mountain mediterranean. 1800–1950 m. Humid, Sub-humid; very cold winter. Basalt, dolomitic limestone. Low slope, almost always exposed to the north.
  Central Middle-Atlas (Tichoukt, Bakrit).
- *Artemisio mesatlanticae-Genistetum pseudoretamoïdis* Quézel, Barbéro, Benabid, Loisel, and Rivas-Martinez 1992b
  Mesomediterranean. 1400–1900 m. Semi-arid; cold winter. Limestone, dolomite, colluvions.
  Eastern Middle-Atlas (south and western flanks of Tichoukt until the region of Immouzer des Marmoucha).
- *Avenastro filiifolii-Erinaceetum anthyllidis* Quézel, Barbéro, Benabid, Loisel, and Rivas-Martinez 1992b
  Mountain mediterranean and Oromediterranean. 2400–2650 m. Sub-humid; cold to very cold winter. Dolomitic limestone.
  Eastern Middle-Atlas (Jbel Tichoukt).
- *Avenastro jahandiezii-Marrubietum ayardii* Quézel 1957
  Oromediterranean. 2100–2700. Semi-arid; cold to very cold winter. Limestone scree; rocky soil.
  Eastern High-Atlas (jbel Ayachi, jbel Maasker), eastern Middle-Atlas (jbel Bou Iblane, jbel Moussa-ou-Salah, jbel Bou-Nacer).

- *Carduncello mesatlantici-Erinaceetum anthyllidis* Nègre 1961
  Mountain mediterranean. 2000–2300 m. Sub-humid; very cold winter. Limestone.
  Central Middle-Atlas (jbel Hayyan, Assaka-n-Aouam, Aari Inifif, col du Zad, jbel Saa).
- *Centaureo triumfetii-Cytisetum balansae* Benabid 1988
  (*Cerastio gibraltarici-Cytisetum balansae* Benabid 1988)
  Mountain mediterranean. 1980–2130 m. Sub-humid; cold to very cold winter. Limestone.
  Central Middle-Atlas (regions of col du Zad, Timahdite, Bou Angar, jbel Koubbate).
- *Centaureo boissieri-Arenarietum armerinae* Nègre 1961
  Mountain mediterranean. Sub-humid; cold to very cold winter. Dolomite, limestone.
  Central Middle-Atlas (jbel Saa).
  Subassociations:

  - *avenetosum filifoliae*
  - *scorzoneretosum pygmae*

- *Genisto pseudopilosae-Bupleuretum lucidae* Quézel, Barbéro, Benabid, Loisel, and Rivas-Martinez 1992b
  Superior mediterranean. 1500–1620 m. Sub-humid; cool to cold winter. Compact limestone.
  Oriental mountains (Gaâda of Debdou).
  Belong to the serial of *Quercus rotundifolia*, association *Festuco scaberrimae-Quercetum rotundifoliae*.
- *Hieracio pseudopilosellae-Adenocarpetum boudyii* Benabid 1988
  Superior mediterranean. 1620–1760 m. Humid; cold winter. Basalt. Deep soil.
  Central Middle-Atlas (region of the Sehb above Azrou) (Figs. 5 and 6).
- *Jasiono humilis-Genistetum quadriflorae* Benabid 1988
  (*Violo dehnhardtii-Genistetum quadriflorae* Benabid 1988)
  Superior mediterranean. 1620–1750 m. Humid; cold winter. Basalt. Brown ferralitic soil.
  Central Middle-Atlas (Bou-Jrirt in the region of Ifrane).
  Association developed in forest clearings.
- *Juniperetum hemisphericae* Quézel, Barbéro, Benabid, Loisel, and Rivas-Martinez 1992b
  Oromediterranean. 2530–2750 m. Sub-humid; cold to very cold winter. Limestone; rocky.
  Eastern High-Atlas (jbel Maâsker, jbel Ayachi), eastern Middle-Atlas (jbel Bou Nacer, jbel Bou Iblane, jbel Tichoukt).
  Citation and informations after Benabid (2000: 275); the authors of this association (Quézel, Barbéro, Benabid, Loisel, and Rivas-Martinez 1992b) did not give any detail, just this sentence (p. 156) "Enfin sur les crêtes apparaissent des peuplements denses à *Juniperus communis subsp. hemisphaerica*".

Fig. 5 *Hieracio pseudopilosae-Adenocarpetum boudy*, Middle-Atlas (ph. Taleb)

Fig. 6 *Adenocarpus Boudyi* (ph. Taleb)

- *Marrubio echinati-Astracanthetum maroccani* Quézel, Barbéro, Benabid, Loisel, and Rivas-Martinez 1992b
  Mountain mediterranean. 1700–2300 m. Sub-humid, Humid; cold to very cold winter. Limestone.
  Eastern Middle-Atlas (north flank of Bou Iblane).
- *Ranunculo millefoliati-Cytisetum balansae* Nègre 1961
  Mountain mediterranean. 1950–2050 m. Sub-humid; very cold winter. Basalts.
  Central Middle-Atlas (basin of Bou Angar between Timahdite and col du Zad).
- *Scabioso tomentosae-Thymetum algeriensis* Benabid 1988
  Supramediterranean. 1800–1900 m. Sub-humid; cold winter. Marly limestone.
  Central Middle-Atlas (Ajgou)
  Top of the supramediterranean mesoxerophile serial of the green oak.
  Subassociations:

  – *phlomidetosum herba-venti*
  – *brometosum erecti*

- *Scutellario demnatensis-Astragaletum numidici* Quézel, Barbéro, Benabid, Loisel, and Rivas-Martinez 1992b
  Superior and Mountain mediterranean. 1550–2100 m. Sub-humid; cold winter. Limestone, flyschs; eroded soils.
  Eastern Middle-Atlas (north flank of Bou Iblane).
  Supramediterranean serial of the green oak (*Genisto jahandiezii-Quercetum rotundifoliae* Barbéro, Quézel, and Rivas-Martinez 1981).
- *Thymo atlantici-Erinaceetum anthyllidis* Quézel, Barbéro, Benabid, Loisel, and Rivas-Martinez 1992b
  Oromediterranean. 2200–2650 m. Semi-arid; cold to very cold winter. Limestone; rocky soils.
  Eastern Middle-Atlas (windy summits of the jbel Bou Iblane, jbel Bou Nacer, etc.).
- *Veronico roseae-Bupleuretum spinosi* Nègre 1961
  Mountain mediterranean. 1700–2100 m. Sub-humid, Humid; cold to very cold winter. Limestone; soils rich with fine particles.
  Central Middle-Atlas (jbel Saa, Afenourir, jbel hayan, Sidi Mguild).
- *Junipero communi-Prunusetum prostratae* Taleb and Fennane 2009
  Oromediterranean. 2570–2750 m. Sub-humid, cold to very cold winter. Limestone; rocky soils (Fig. 7).
  Eastern High-Atlas (Maâsker, Ayachi), eastern Middle-Atlas (Tichoukt).
- *Arenario armerinae-Sideretum matris-filiae* Romo 2009
  Oromediterranean. 2600–2700 m. Sub-humid; cold winter. Rocky soils.
  Eastern Middle-Atlas (Tichoukt).
- *Association with Thymus zygis* Br.-Bl. and Maire 1924
  Mesomediterranean. 1200–1250 m. Sub-humid; cool to cold winter. Basalt.
  Central Middle-Atlas.
  Association belonging to the green oak serial.

Fig. 7 *Junipero communi-Prunusetum prostratae*, Eastern High-Atlas (ph. Taleb)

Ormenion scariosae Quézel 1952

- *Adenocarpo anagyrifolii-Genistetum maroccanae* Quézel 1957

  Superior- and Mountain mediterranean. 1800–2400 m. Semi-arid, Sub-humid; cold winter. Schists.

  High-Atlas (Seksaoua, Rherhaya, south east flank of the Oukaimeden, Tizi n'Tichka, Aremd, Afensou, north-eastern flank of the jbel Tizerag, Tizi n'Tichka road near Immouzer).

  Subassociations:

  – *subassociation typicum*
  – *bupleuro spinosae-stipetosum nitensis*

- *Cedro atlanticae-Helianthemetum crocei* Nègre 1952

  Mountain mediterranean. 2100–2300 m. Humid, Sub-humid; cold winter. Dolomite, sandy clay, dolomitic sand.

  Central Middle-Atlas (jbel Saa).

  Subassociations:

  – *helianthemetosum crocei*
  – *quercetosum ilicis*

- *Genisto myrianthae-Bupleuretum atlantici* Quézel 1957

  Superior and Mountain mediterranean. 2000–2500 m. Semi-arid; cold winter. Limestone, rocky slopes.

  High-Atlas (Andromer, Rhat, Mgoun, Azourki, Aioui, Imrhas, Maasker, the lakes plateau).

  Subassociations:

  – *bupleuro-stipetosum nitensis*
  – *cistetosum laurifolii*
  – *thymetosum satureioidis*

- *Ononido atlantici-Santolinetum canescentis* Quézel 1957

  Superior and Mountain mediterranean. 2000–2500 m. Semi-arid; cold winter. Calcareous schist.

  Western High-Atlas (Seksaoua mountains in the region of Ras Moulay Ali, Aoulime, Timerguet...).

  Degradation groupement of the green oak forest.

- *Ormenido scariosae-Euphorbietum megalatlanticae* Taleb and Fennane 2003

  Superior and Mountain mediterranean. 1900–2200 m. Semi-arid, Sub-humid; cold winter. Limestone; stony soils.

  Eastern High-Atlas (the lakes plateau, southern slopes of the jbel Msadrid, very locally in the clearings of the green oak forest on the southeast slope of the jbel Hayim).

- *Salvio aucherii-Avenetum filifoliae* Quézel 1957

  Superior and Mountain mediterranean. 1700–2600 m. Semi-arid, Sub-humid; cold winter. Limestone, dolomitic limestone, marly calcareous; rocky soils.

  Eastern Middle-Atlas (eastern flank of the jbel Bou Nacer).

- *Teucrio malenconiani-Thymetum brevidentis* Quézel, Barbéro, Benabid, and Rivas-Martinez 1994

  Mountain mediterranean. Semi-arid; cold to very cold winter. 2200–2470 m. Granite, quartzite.

  Anti-Atlas (Summits of Saghro in Tiouit, jbel Feggour above Iknioun, Amalou n'Mansour).

- *Bupleuro spinosi-Achilleetum odoratae* Peltier 1982

  Mountain mediterranean. Semi-arid; cold to very cold winter. 2100–2450 m. Loamy-argillaceous.

  High siliceous summits of the center and east of the Anti-Atlas: jbels Aklim, Imguend and Iguiguil.

- *Ormenio scariosae-Marrubietum litardieri* Peltier 1982

  Superior or? Mountain mediterranean. 1900–2200 m. Semi-arid; cold winter. Quartzites, conglomerate; silts or clayey silts, brown color.

  Anti-Atlas (jbels Aklim and Imguend).

Arenarion pungentis Quézel 1957

- *Agropyro festucoidis-Scrophularietum macrorrhynchae* Quézel 1957
  Mountain mediterranean, Oromediterranean. 2150–3000 m. Sub-humid; very cold winter.
  High-Atlas (Andromer, Rhat, Atlas of Zaouiat Ahansal, Mgoun, jbel Aioui), eastern High-Atlas (jbel Ayachi, jbel Maâsker, jbel Aberdouz), eastern Middle-Atlas (jbel Bou Iblane, jbel Bou Nacer) (Fig. 8).
  The association growths in a more humid habitats than the general climate of atlasic formations of spiny xerophytes. Recently seen by Taleb on the north flanks of Bou Iblane in years 2011 and 2016 and Bou Nacer in 2016.
  Subassociations:
  – *subassociation with Festuca mairei*
  – *subassociation with Agropyron festucoides, Artemisia atlantica and Marrubium atlanticum*

- *Arenario pungentis-Vicietum rerayensis* Quézel 1957
  Mountain mediterranean and Oromediterranean. 2400–3400 m. Semi-arid, Sub-humid. Siliceous, granite, rhyolite.
  Central and western High-Atlas (jbel Tamgout, jbel Tichka, Reraya valley, jbel Bou Ourioul, Oukaimeden, Tazerarht, Toubkal, Seksaoua).

Fig. 8 *Agropyro festucoidis-Scrophularietum macrorrhynchae*, Eastern Middle-Atlas (Bou-Nacer) (ph. Taleb)

- *Elyzaldio violaceae-Spergularietum flaccidae* Quézel 1957

  Oromediterranean. 2700–3050 m. Semi-arid; very cold winter. Granitic sands. Western High-Atlas (Tichka plateau in the Seksaoua).

  Transition association between dry pozzines of *Silene Thomsoni* + *Potentilla pennsylvannica* and the spiny xerophytic garrigue.
- *Teucrio musimoni-Avenetum montanae* Quézel 1952

  Oromediterranean. 2900–3450 m. Semi-arid; very cold winter. Limestone; rocky substrate.

  High-Atlas (jbel Ayachi, jbel Rhat), Eastern Middle-Atlas (Bou Nacer, Guelb er-Rahal).
- *Velletum mairei* Quézel 1952

  Mountain mediterranean and Oromediterranean. 2250–3000 m. Semi-arid, Sub-humid; very cold winter. Limestone, dolomite calcareous; rocky soils.

  Eastern Middle-Atlas (jbel Bou Nacer, jbel Bou Iblane) High-Atlas (jbel Ayachi, jbel Maâsker, jbel Msadrid, jbel Fazzaz, jbel Sloul, the lakes plateau, jbel Andromeur, jbel Azourki, jbel Rhat, jbel Mgoun, Imdrhas, Adrar m'Korn, jbel Aori, Ahansal) (Figs. 9 and 10).

  Subassociations:

  – *alyssetosum spinosae*
  – *erinaceetosum anthyllidis*
  – *pterocephaletosum depressi*

Fig. 9 *Velletum mairei*, Eastern Middle-Atlas (ph. Taleb)

Fig. 10 *Vella mairei* (ph. Taleb)

- *Groupement* of *Scorzonera pygmaea* and *Catananche coespitosa* Taleb and Fennane 2003
  Mountain mediterranean, Oromediterranean. Semi-arid; cold winter. 2200–2500 m. Indifferent to substrate; earthy soil.
  Eastern High-Atlas.

Platycapnion saxicolae Quézel 1952

- *Cirsio dyris-Silenetum ayachicae* Quézel 1952
  Oromediterranean. 2500–3600 m. Semi-arid, Sub-humid; very cold winter. Limestone; moving rocks.
  High-Atlas (jbel Ayachi, jbel Rhat, jbel Mgoun, jbel Tichka, jbel m'Korn, jbel Aori, Tazerharht), Eastern Middle-Atlas (jbel Bou Nacer, jbel Bou Iblane).
  Endemic of the High-Atlasic summits, especially on the screes at the foot of the cliffs where *Saxifraguetum gausseni* clings. Represented on the siliceous High-Atlas by a subassociation with *Cicer atlanticum*.
- *Raffenaldio primulinae-Campanuletum anremericae* Quézel 1957
  Oromediterranean. 3350–3650 m. Semi-arid; very cold winter. Limestone; small moving stones.
  High-Atlas (jbel Ayachi, jbel Mgoun, jbel Rhat, jbel Aori).
- *Violo calcareae-Vicietum anremericae* Quézel 1957
  Oromediterranean. 3450–3600 m. Semi-arid; very cold winter. Limestone.
  High-Atlas (Andromer).

- *Violo dyris-Linarietum luridae* Quézel 1957
  Oromediterranean. Above 3400 m. Semi-arid; very cold winter.
  Central High-Atlas (Erdouz, Toubkal massif, Bou Ourioul).
- *Violetum orientalis* Quézel 1957
  Oromediterranean. 3400–3700 m. Semi-arid; very cold winter. Limestone.
  High-Atlas (jbel Ayachi, jbel Rhat, jbel Mgoun).
  Subassociations:

  – *veronicetosum chartonii*
  – *asperuletosum litardieri*

Lawns, Pozzines and Meadows of High Mountains

Poetea bulbosae Rivas Goday and Rivas-Martinez in Rivas-Martinez 1978

Seasonal, perennial and ephemeral pastures. Humid, Sub-humid, Semi-arid. Meso- to Oromediterranean belts.

Poetalia bulbosae *Rivas Goday and Rivas-Mart. in Rivas Goday and Ladero 1970*

Potentillo maurae-Campanulion filicaulis Galan de Mera and Vicente Orellana 1997

(*Trisetario flavescentis-Caricion chaetophyllae* Benabid 1988)

- *Caro jahandiezii-Potentilletum maurae* Nègre 1961
  (*Poo-Caricetum divisae* Nègre 1961)
  Superior, Moutain mediterranean. 1800–2300 m. Sub-humid; cold winter. Limestone, basalt. Silty, clayey, deep soil.
  Middle-Atlas (jbel Saa).
- *Phleo nodosi-Poetum bulbosae* Benabid 1988
  Superior, Mountain mediterranean. 1800–1850 m. Sub-humid; cold winter. Limestone; clayey soil, colluvions.
  Central Middle-Atlas (Afennourir, limestone causse of Aïn Leuh).
- *Sileno mekinense-Caricetum chaetophyllae* Benabid 1988
  Superior Mediterranean. 1600–1700 m. Sub-humid; cold winter. Limestone; more or less rocky.
  Central Midle-Atlas (Tissfoula, Iguer Aouragh).
  Degradation groupement of the *Arenario armerinae-Adenocarpetum boudyi*.

M. S. Taleb, M. Fennane, *Vascular Plant Communities of Morocco*, Geobotany Studies, https://doi.org/10.1007/978-3-319-93704-5_3

Fig. 1 Pozzines, National Park Oriental High-Atlas (ph. Taleb)

- *Vulpio genuinae-Limonietum moureti* Benabid 1988
  Superior mediterranean. Humid; cold winter. 1500–1650 m. Limestone; basic vertisols, wet in winter, dry in summer.
  Central Middle-Atlas (depressions of the causse of Aïn Leuh).

Other Groupements

- *Drabo lutescentis-Violetum minimae* Romo 1988
  Oromediterranean. 2300–2400 m. Sub-humid; very cold winter.
  Central High-Atlas (Okaimeden).
  Thero-nitrophilous association.

Nardetea strictae Rivas Goday ex Rivas Goday and Rivas-Martinez 1963

Grazed acidophilous grassland. Sub-humid, Semi-arid; very cold winter. Acid organic mineral soils. Oromediterranean belt (Figs. 1 and 2).

Fig. 2 Pozzines, Eastern High-Atlas (ph. Taleb)

Nardetalia strictae *Oberdorfer ex Preising 1949*

Trifolion humilis Quézel 1952

- *Festuco rubrae-Gentianetum ciliatae* Quézel 1957

  Oromediterranean. 2450–3350 m. Semi-arid, Sub-humid; very cold winter. Limestone. Clayey, loamy, sandy.

  Lawns, Pozzines and meadows of high mountains of the central High-Atlas (Mgoun, Tessaout, Andromer, Aïoui, Azourki).
- *Festuco hystrix-Taraxacetum atlanticae* Quézel 1952

  Oromediterranean. 2400–3300 m. Semi-arid, Sub-humid; very cold winter. Limestone, coarse sand, loam, clay. Stony soils.

  Lawns, Pozzines and meadows of high mountains of the eastern High-Atlas (Ayachi), central High-Atlas (Mgoun, Rhat, Azourki, Aïoui) and eastern Middle-Atlas (Bou Iblane, Moussa ou-Salah, Bou-Nacer).

  Subassociations (Quézel 1957: 290):

  – *subassociation typicum*
  – *rorippetosum hayanicae*
  – *festucetosum maroccanae*

- *Gentiano tenellae-Agrostidetum rupestris* Quézel 1957

  Oromediterranean. 3000–3600 m. Semi-arid, Sub-humid; very cold winter. Humus, coarse sand, fine sand, silt, clay.

Lawns, Pozzines and meadows of high mountains of the central High-Atlas (Toubkal and high valley Rherhaya).

- *Nardo strictae-Festucetum yvesiani* Quézel 1957

Oromediterranean. 2400–3200 m. Semi-arid, Sub-humid; very cold winter. Granite, rhyolite, andesite, humus, coarse sand, fine sand, clay, silt.

Lawns, Pozzines and meadows of high mountains of the central and western High-Atlas (massif Tichka, Bou Ourioul, High Reraya, Isougane Ouagouns, Oukaïmeden: Edge of the brook descending from Angour).

- *Nardo strictae-Gentianetum atlanticae* (Maire 1924) Quézel 1952

Oromediterranean. Semi-arid, Sub-humid; very cold winter.

Lawns, Pozzines and meadows of high mountains of the High-Atlas.

- *Poo alpini-Anacycletum atlantici* Quézel 1957

Oromediterranean. 3600–3900 m. Semi-arid, Sub-humid; very cold winter. Acid eruptive rocks; earthy soils.

Lawns, pozzines and meadows of high mountains of the central High-Atlas (massif of Toukal).

Subassociations:

– *pooetosum alpini*
– *anacylusetosum atlantici*

- *Potentillo hispanicae-Silenetum thomsonii* Quézel 1957

Oromediterranean. 2400–3200 m. Semi-arid, Sub-humid; very cold winter. Humus, pebbles, coarse sand, fine sand, silt, clay.

Lawns, pozzines and meadows of high mountains of the central High-Atlas (from Atlas of Seksaoua until Tizi n'Tichka: massif Tichka, massif Toubkal, Agdals of the Oukaimeden and Bou Ourioul).

Subassociations:

– *subassociation typicum*
– *festucetosum marocanae*

- *Potentillo tornezianae-Rorippetum atlanticae* Quézel 1957

Oromediterranean. 2600–3900 m. Semi-arid, Sub-humid; very cold winter. Sciliceous; stony, wet soils.

Lawns, pozzines and meadows of high mountains of the central and western High-Atlas (massif of Toubkal, plateau of Tichka, High Reraya, Bou Ourioul, north flank of Siroua, north flank of tazerharht, Azourki).

- *Raffenaldietum platycarpae* Quézel 1952

Oromediterranean. 3100–3800 m. Semi-arid, Sub-humid; very cold winter. Clayey.

Central and eastern High-Atlas, Estern Middle-Atlas.

Lawns, pozzines and meadows of high mountains.

Subassociations:

– *subassociation typicum* (central High-Atlas: jbel Rhat)

– *subassociation glaciale*: eastern High-Atlas (Ayachi), central High-Atlas (Mgoun, Rhat), eastern Middle-Atlas (Bou Nacer, Bou Iblane).

• *Trifolio repensae-Deschampsietum coespitosae* Quézel 1957
Oromediterranean. 2600–3000 m. Semi-arid, Sub-humid; very cold winter. Granitic sand, more or less humified.
Lawns, pozzines and meadows of high mountains of the western and central High-Atlas (massif of Tichka, Agdal of the Oukaimeden, lower pozzines of the sources of the Tessaout) Anti-Atlas (north flank of the Siroua).

Mulgedio-Aconitetea Hadac et Klika in Klika et Hadac 1944

(*Betulo-Adenostyletea* Br.-Bl. 1948 p.p.)
Perennial tall herbaceous communities at high altitudes. Humid, Sub-humid; very cold winter. Wet fertile soils. Oromediterranean, Mountain mediterranean.

Adenostyletalia alliariae *Br.-Bl. 1930*

(*Cirsietalia flavispinae* Quézel 1957)

Cirsion flavispinae Quézel 1953

• *Cirsio chrysanthemi-Heracleetum montanii* Quézel 1957
Mountain mediterranean, Oromediterranean. 2000–3600 m. Sub-humid; very cold winter. Siliceous, sandstone.
Central High-Atlas (between Tizi Machou and du Tizi n'Tichka).
Subassociations:

– *subassociation typicum*
– *cirsietosum chrysacanthemi*
– *festucetosum mairei*
– *eryngietosum variifolii*

• *Cirsio flavispini-Heracleetum embergeri* Quézel 1957
Mountain mediterranean, Oromediterranean. 1800–3400 m. Sub-humid, Humid; very cold winter. Limestone. Stony.
High-Atlas (Ayachi, Rhat, north flank of jbel Maasker, jbel Aioui, Akka n'Tazzert, Imi n'Ouaka, Tizi n'Tichka, Zaouiat Ahansal).

Subassociations:

- *subassociation typicum*
- *aconietosum neapolitani*
- *eryngietosum variifolii*

Eryngion variifolii Quézel 1957

- *Bellis caerulescendis-Heracleum sphondyllii* Romo 2009
  Oromediterranean. 2650–2710 m. Sub-humid; very cold winter.
  High-Atlas (Adrar angour, Oukaïmeden).
- *Cephalario maroccanae-Inuletum maletii* Romo 2009
  Mountain mediterranean. 1800–1900 m. Humid; cold winter.
  Central Middle-Atlas (Michliffen).

Scheuchzerio palustris-Caricetea fuscae Tüxen 1937

Humid substrates, weeping, mainly siliceous. Vegetation dominated by *Cyperaceae* species. Oromediterranen, Mountain mediterranean.

Caricetalia fuscae *(Koch 1926) Nordhagen 1937 em. Br.-Bl. 1949*

Caricion intricatae Quézel 1957

- *Poo rivularis-Veronicetum repensis* Quézel 1957
  Mountain mediterranean, Oromediterranean. 1900–3200 m. Sub-humid; very cold winter. Rhyolite, granites, marl, compact limestone.
  High-Atlas (Ayachi, lake plateaux, jbel Rhat, Bou Ourioul, jbel Tichka, High Reraya, Tizi n'Tichka, Isougane n'Ouagouns).
  Subassociations:
  - *subassociation typicum*
  - *heliosciadetosum repensis*
- *Schoeno nigrici-Triglochinetum palustris* Quézel 1957
  Oromediterranean. 2500 m. Sub-humid; very cold winter. Compact limestone; dark humus.
  Central High-Atlas (jbel M'Goun, Jbel Andromeur, Tessaout gorges).

Montio-Cardaminetea Br.-Bl. and Tüxen ex Klika and Hadac 1944

Edges of streams and water sources.

Montio-Cardaminetalia *Pawlowski, Sokolowski and Wallish 1928*

Cardamino-Montion Br.-Bl. 1926

- *Sagino sabuletori-Stellarietum uliginosae* Quézel 1957

  Mountain mediterranean, Oromediterranean. 2300–3300 m. Sub-humid; very cold winter. Granites, rhyolite, marly calcareous.

  Western and central High-Atlas (jbel Tichka, High Reraya near the Nelter refuge, Azourki,Tazerarht, Bou Ourioul).

  Seems to represent the atlasic vicariance of associations with *Bryum schleicheri* of the European mountains.

  Subassociations:

  – *subassociation with Ranunculus fluitans, Montia rivularis and Philonotis seriata*
  – *subassociation with Sagina sabuletorum var. atlantica and Epilobium atlanticum*

Other Groupements

- *Eryngio maroccanae-Orchidetum latifoliae* Nègre 1961

  Mesomediterranean, Superior mediterranean. Sub-humid; cold to very cold winter.

  Central Middle-Atlas (region of jbel Saa).

  High meadow, always green, completely covering the ground

  Subassociations:

  – *plantagonetosum (lanceolatae) lacustris*
  – *subassociation with Festuca arundinacea, Eryngium maroccanum or Orchis coriophora*

Rupicolous Communities

Adiantetea capilli-veneris Br.-Bl. in Braun-Blanquet, Roussine and Nègre 1952a

Chasmophytic communities, rich in bryophytes, on shaded cliffs and walls, in halve-caves with dripping water and in water-splashed habitats.

Adiantetalia capilli-veneris *Br.-Bl. ex Horvatic 1934*

Adiantion capilli-veneris Br.-Bl. ex Horvatic 1934

- *Adianto-Hypericetum naudiniani* Deil 1996

  Mesomediterranean. 1200–1500 m. Humid, Sub-humid; cool to cold winter.

  Central Rif (south flank of the jbel Arz between Mazouz and Targhzout), Middle-Atlas.

  Vertical cliffs on schists, covered by calcareous tuffs (*subassociation typicum*); permanent streams near sources (*philonotidetosum fontanae*).

  Subassociations:

  - *subassociation typicum*
  - *philonotidetosum fontanae*

- *Adianto-Hypericetum pubescentis* Varo Alcala and Fernandez Casas 1970

  Mesomediterranean. 900–1600 m. Sub-humid, Humid; cool to cold winter. Basalt.

  Middle-Atlas (Oued Srou), Rif.

  Dripping water (*subassociation typicum*); running water (*cratoneuretosum filicini*).

M. S. Taleb, M. Fennane, *Vascular Plant Communities of Morocco*, Geobotany Studies, https://doi.org/10.1007/978-3-319-93704-5_4

Subassociations:

- *subassociation typicum*
- *cratoneuretosum filicini* Deil 1996

- *Samolo valerandi-Adiantetum capilli-veneris* Julve ex de Foucault 2015
  (*Eucladio verticillati-Adiantetum capilli-veneris* sensu Br.-Bl. et al. 1952a)
  Thermo- and Mesomediterranean. 0–1300 m. Sub-humid, Humid; warm to temperate winter.
  Widespread in Atlantic Moroccan, Rif and mountains (Deil, pers. comm.).
- *Trachelio coerulei-Adiantetum capilli-veneris* Bolos 1957b
  Thermomediterranean, Mesomediterranean. Sub-humid; cool winter. Limestone, carbonated clays.
  Widespread in Atlantic Moroccan and mountains (Deil, pers. comm.).
  Rupicolous formations on cliffs, walls and along water-courses.
  Subassociations:

  - *pteridietosum vittatae*
  - *hypericetosum metroi* Deil 1996

Asplenietea trichomanis (Br.-Bl. in Meier and Braun-Blanquet 1934) Oberdorfer 1977

Chasmophytic communities of crevices, rocky ledges, rocky cliffs and walls.

Potentilletalia caulescentis *Br.-Bl. in Braun-Blanquet and Jenny 1926*

Potentillion caulescentis Br.-Bl. in Braun-Blanquet and Jenny 1926

- *Centaureo benoistii-Galietum ephedroidis* Romo 2008
  Mesomediterranean. 1300–1600 m. Semi-arid, Arid; cool to cold winter. Limestone.
  Saharan Atlas, Eastern High-Atlas (cliffs of the Ziz river, above Aït Serhouchène, toward the Rich, from jbel Tazigzaout in the north until jbel Izouggarn in the south).
  Rupicolous association, in the area of *Adenocarpo bacquei-Buxetum balearicae*.

Rupicapnion africanae Daumas, Quézel, and Santa 1952

- *Rupicapnetum mairei* Br.-Bl. and Maire 1924
  (*Association with Fumaria africana* Br.-Bl. and Maire 1924)
  (*Groupement with Rupicapnos decipiens* Deil 1994) [Deil and Galan de Mera 1996: 106]
  (*Rupicapnetum africanae* Perez Latorre, Cabuzedo, and Nieto 1995)
  Mesomediterranean, upper Thermomediterranean. 900–1300 m. Sub-humid; cold to cool winter. Cracks in calcareous and dolomitic rocks, limestone cliffs and overhangs.
  Central High-Atlas (region of Demnate), Middle-Atlas (region of El Hajeb, region of Azrou), Northern atlantic Morocco (Pre-Rif), Eastern High plateaux, Rif. [Deil and Galan de Mera 1996: 102].

Violion saxifragae Quézel 1952

- *Anthyllidetum warnieri* Quézel 1952
  Superior mediterranean, Oromediterranean. 1900–2400 m. Sub-humid; cold to very cold winter. Limestone. Rocks.
  Central High-Atlas (Imi-n-Ouaka gorges, north flank of jbel Azegza, north flank of jbel Alimane, Timeourine gorges, Mgoun: Akka n'Aït Ahmet gorges, Assif Isladen gorges, Imi-n-Ouaka, gorges between jbel Azegza and jbel Alimane).
- *Arabido alpinae-Festucetum atlanticae* Romo 2008
  Oromediterranean. 2600–3100 m. Semi-arid, Sub-humid; very cold winter. Granite. Cliffs, rocky outcrops.
  High-Atlas (from jbel Ousdim and plateau Tichka until jbel Angour and massif Oukaimeden).
- *Arenarietum mairei* Quézel 1952
  Oromediterranean. Above 2700 m. Semi-arid, Sub-humid; very cold winter.
  Central High-Atlas (Mgoun, Rhat).
- *Campanulo embergeri-Erodietum atlanticae* Quézel 1957
  Mountain mediterranean, Oromediterranean. 1800–2500 m. Semi-arid, Sub-humid; cold to very cold winter. Schists; rocks.
  Eastern and central High-Atlas (Seksaoua valleys, west flank of Ras Moulay Ali, rocks above Assif Lalou, Tazerhart massif, rocks of Tizi m'Zic west of Aremd, north-western flank of jbel Siroua).
- *Drabetum oreadae* (Maire 1924) Quézel 1957
  Oromediterranean. From 3500 m and above. Semi-arid; very cold winter. Limestone, rocks.
  Eastern High-Atlas (Ayachi massif).
- *Drabo hispanicae-Pterocephaletum kunkeliani* Romo 2008
  Mountain mediterranean. 1800–1980 m. Sub-humid; cool winter. Limestone.

Eastern Rif (cliffs of the jbel Azrou Akechar, above Tizi-Ouzli).

Rupicolous association, in the *Quercus rotundifolia* forest.

- *Gnaphalietum genevoisii* Quézel 1952

  Oromediterranean. Around 3200 m. Semi-arid; very cold winter.

  Central High-Atlas (Rhat, Mgoun).

- *Sarcocapnetum atlanticae* Quézel 1952

  Oromediterranean. 2500–3200 m. Sub-humid, Semi-arid; very cold winter.

  High-Atlas: central gorges of the Rhat, rocky outcrops exposed to the north in the Rhat, rocky slopes exposed to the north in the Mgoun, north flank of the jbel Aori, gorges of Akka n'Tazzert (jbel aioui), upper gorges of Tessaout.

  According to Deil (pers. comm.), this association belongs to *Sarcocapnion enneaphyllae*.

- *Sarcocapno crassifoliae-Erodietum tordylioidis* Daumas, Quézel, and Santa 1952

  (Association with *Geranium cataractarum* Maire 1924) [Deil and Galan de Mera 1996:106]

  Mesomediterranean, Mountain mediterranean. Sub-humid, Semi-arid; cool to cold winter.

  Middle-Atlas (Bni Mellal, Aghbala, El Ksiba, Guigou, Timahdite) [Deil and Galan de Mera 1996:106].

  According to Deil (pers. comm.), this association belongs to *Sarcocapnion enneaphyllae*.

- *Saxifragetum gausseni* Quézel 1952

  (*Biscutelletum frutescentis ayachicum* Quézel 1952) (Deil and Galan de Mera 1996:105)

  Oromediterranean. 3000–3500 m.Semi-arid, Sub-humid; very cold winter. Limestone; rocks.

  Central High-Atlas (Mgoun, Rhat, Ayyachi).

- *Saxifrago maireanae-Phagnaletum platyphyllae* Quézel 1957

  Mountain mediterranean, Oromediterranean. 2000–3000 m. Semi-arid, Sub-humid; cold to very cold winter. Schists, sandstone; rocks.

  Central High-Atlas (western flank of Bou Ourioul, Oukaimeden (jbel Tizerarg), Sidi Chamharouch, Toubkal massif, Tazerhart massif, Azibs Tamzikht).

  Subassociations:

  – *subassociation with Saxifraga maireana and Sarcocapnos crassifolia var. subspeciosa:* northern exposure
  – *subassociation with Sempervivum atlanticum and Centaurea incana var. fulgida:* southern exposure

- *Saxifrago globuliferae-Seslerietum argentae* Romo 2008

  Mesomediterranean. 1400–1600 m. Sub-humid, Humid; cool winter. Limestone.

  Western Rif (cliffs of jbel Lakraa, jbel Tissouka, Sfiha Telj).

  Rupicolous association, in the *Quercus rotundifolia* forest and also with *Abies maroccana*.

- *Saxifrago tricrenatae-Poetum ligulatae* Romo 2008
  Mesomediterranean, Mountain mediterranean. 1600–1900 m. Humid; cool to cold winter. Limestone cliffs.
  Western Rif (above Chefchaouene, jbel Assillenh, valley Tissin-Lel, massif Tissouka, jbel Lakraa).
- *Sileno dyris-Silenetum boryi* Quézel 1957
  Oromediterranean. 2500–3400 m. Semi-arid, Sub-humid; cold to very cold winter. Granite, schists, limestone; rocks.
  Western and central High-Atlas (west flank of Ras Moulay Ali, High valley of Seksaoua, Rherhaya near Sidi Chamharouch, Aremd, rocks of Tizi n'Ousden, north and eastern flanks of Tazerhart and Isougane Ouagouns, east flank of agdal Bouidarène, north and western flanks of Tamguerd Imadène).
  Subassociations:

– *subassocation typicum*
– *subassociantion occidentale*
– *subassociation with Arabis erubescens*

Sarcocapnetalia enneaphyllae *Fernandez Casas 1972*

(*Coeno-Sarcocapnetalia* Deil and Galan de Mera 1996)

Sarcocapnion enneaphyllae Fernandez Casas 1972

(*Coeno-Sarcocapnion* Deil and Galan de Mera 1996)

- *Saxifrago globuliferae-Geranietum cataractari* Romo 2008
  Mesomediterranean, Mountain mediterranean. 1500–2000 m. Sub-humid Humid; cool to cold winter. Consolidated limestone cliffs.
  Middle-Atlas, on the flanks exposed directly to western atlantic influences.
- *Sarcocapnetum enneaphyllae* Rivas Goday 1941
  Thermomediterranean. Semi-arid; warm to temperate winter. Limestone; rocky.
  Mediterranean coast (Nekor-Bokkoya) (Deil and Galan de Mera 1996; Deil and Hammoumi 1997).

Tinguarretalia siculae *(Daumas, Quézel, and Santa 1952) Galan de Mera in Pérez Latorre et al. 1996*

Campanulion velutinae Martinez-Parras and Peinado 1990

- *Soncho fragilis-Rhodanthemetum laouense* Deil 1994

Thermomediterranean.250–600 m. Semi-arid, Sub-humid; temperate winter. Limestone.

Rif (Oued Laou valley in Talembote).

- *Stachyo circinatae-Rhodanthemetum hosmariense* Deil 1994

  Thermomediterranean, lower Mesomediterranean. Semi-arid, Sub-humid; temperate winter. Limestone.

  Rif (from jbel Moussa until south of Tetouan).

Celsio antiatlanticae-Aeonion arborescentis Médail and Quézel 1999

- *Jasiono antiatlanticae-Saturejetum arganietorum* Médail and Quézel 1999

  Infra-, Thermomediterranean. Semi-arid; cool winter. 650–1200 m. Limestone, volcanic substrate.

  Anti-Atlas (regions of Tighmi and Tafraoute).
- *Ononido angustissimae-Helianthemetum confertae* Peltier 1982

  (*Ononidetum angustissimae* Br.-Bl. and Maire 1924)

  Inframediterranean. 50–100 m. Arid, Semi-arid; warm winter. Sandy.

  Middle atlantic Morocco (regions of Agadir and Essaouira).
- *Scillo latifoliae-Saturejetum monanthae* Médail and Quézel 1999

  Inframediterranean. 300–450 m. Saharan, Arid; temperate winter. Quazrtzite, sandstone.

  Gorges of Kheneg el Hamman, Southern-west of Goulimine, near the road to El Ayoun du Draa, gorges of Oued Noun (Maire ined. 1937), gorges of Oued Ouarsik, South-western of Ifni.
- *Teucrietum tananici* Médail and Quézel 1999

  Thermomediterranean. 700–750 m. Semi-arid; temperate to cool winter. Limestone, rocky substrate.

  Western High-Atlas (region of Immouzzer Ida-ou-Tanane).

Feerion angustifoliae Deil and Galan de Mera 1996

- *Nivelleo nivellei-Feerietum angustifoliae* Deil and Galan de Mera 1996

  (*Rupicapnetum africane trachelietosum* sensu Braun-Blanquet and Maire 1924) [Deil and Galan de Mera 1996: 106]

  Meso- and Superior mediterranean. Sub-humid; cool to cold winter. Cliffs.

  Middle-Atlas (Demnate, vertical cliffs near Sidi Yahya Sawad in the south of Al Qbab) (Braun Blanquet and Maire 1924; Deil and Galan de Mera 1996: 102).

Poterion ancistroidis Br.-Bl. in Meier and Br.-Bl. 1934 em. Deil and Galan de Mera 1996

- *Arenarietum dyris* Quézel 1952

  Oromediterranean. 2900–3400 m. Semi-arid; cold winter. Limestone, cliffs.

Eastern High-Atlas (jbel Ayachi, jbel Maasker).

According to Deil (pers. comm.), this association belongs to *Violion saxifragae.*

- *Drabetum mariae-aliciae* (Quézel 1952) Deil and Galan de Mera 1996

  Oromediterranean. From 3500 m and above. Semi-arid; very cold winter. Limestone, rocks.

  Eastern High-Atlas (jbel Ayachi).

  According to Deil (pers. comm.), this association belongs to *Violion saxifragae.*
- *Erigeronto celerieri-Valerianetum globulariifolii* Quézel 1957

  Oromediterranean. Above 3000 m. Semi-arid; very cold winter. Limestone.

  Eastern Middle-Atlas (Bou-Nacer, Bou Iblane, glacial circus on the western flank of Moussa-ou-Salah).

  According to Deil (pers. comm.), this association belongs to *Violion saxifragae.*
- *Pitardietum coerulescentis* Quézel 1957

  Superior mediterranean, Oromediterranean. 1800–2300 m. Limestone. Rocks.

  Eastern Middle-Atlas (south eastern flank of Bou-Nacer and Guelb er-Rahal).

  According to Deil (pers. comm.), this association belongs to *Violion saxifragae.*
- *Saxifragetum mesatlantici* Quézel 1957

  Superior mediterranean, Oromediterranean. Around 2800 m. Semi-arid, Sub-humid; cold to very cold winter. Limestone, rocky cliffs.

  Eastern Middle-Atlas (Bou-Nacer, Bou Iblane, Tichoukt (Lalla Oum el-Bent)).

  According to Deil (pers. comm.), this association belongs to *Violion saxifragae.*
- *Sedo wilczekiani-Sonchetum masguindalii* Deil and Hammoumi 1997

  Thermomediterranean. 10–100 m. Semi-arid; warm to temperate winter. Limestone, cliffs.

  Mediterranean coast (Bokkoya).

  Subassociations:

  - *subassociation typicum*
  - *silenetosum obtusifoliae*
  - *stipetosum tenacissimae*

- *Campanulo tlemcensis-Teucrietum buxifolii* Daumas et al. 1952

  Limestone cliffs from Tlemcen to Beni Snassen and Debdou.

Other Groupements

- *Bubonio imbricati-Centaurieetum gentilii* Peltier 1982

  Inframediterranean. 30–60 m. Arid; warm winter. Limestone coarse sand.

  Middle atlantic Morocco (National Park of Massa: dunes of Tifnit).
- *Chrysanthemo carinati-Senecietum gallici* Peltier 1982

  Inframediterranean. 10–50 m. Arid; warm winter. Quartzitic coarse sand.

Middle atlantic Morocco (National Park of Massa: between the sea and the dunes of Tifnit).

Anomodonto-Polypodietea Rivas-Martinez 1975

Chomophyte, chasmophyte and epiphytic shaded communities dominated by short rhizomatous ferns ; crevices and surface of rocky cliffs.

Anomodonto-Polypodietalia *de Bolos and Vives in O. de Bolos 1957b*

Polypodion serrati Br.-Bl. in Braun-Blanquet et al. 1952

(*Batramio-Polypodion serrati* O. de Bolos and Vives in O. de Bolos 1957b)

- *Davallio canariensis-Sedetum (hirsuti) baetici* Deil 1994
  Thermomediterranean. 100–800 m. Sub-humid; warm to temperate winter. Sandstone.
  Rif (Tangerian peninsula).
  Subassociation:
  - *silenetosum rosulatae-andryfoliae* Galan de Mera et al. 1996

Selaginello denticulatae-Anogrammion leptophyllae Rivas-Martinez et al. 1999

- *Selaginello denticulatae-Anogrammetum leptophyllae* Molinier 1937
  Rif (Tangerian Peninsula).
- *Pterogonio gracilis-Davallietum canariensis* Pérez Latorre et al. 2000
  Rif (Tangerian Peninsula); epiphytic and schistous screes.

Aquatic (Fresh Water) and Hygrophile Environments

Ecosystems still more or less well present in Morocco, but under very high human pressure (drainage, overgrazing, urbanism, tourism…). They are very disturbed almost anywhere. (Figs. 1, 2, 3, 4, 5 and 6).

Isoëto-Nanojuncetea Br.-Bl. and Tüxen in Braun-Blanquet, Roussine, and Nègre 1952a

Wetland. Herbaceous vegetation—annual and/or perennial—temporarily flooded with freshwater.

Orders and alliances to be studied; at least one order, *Isoëtetalia* Br.-Bl. 1936, seems to be present.

- *Community of Agrostis pourretii* Molina, Tahiri, Agostinelli et al. 2009
  Thermomediterranean. Semi-arid, Sub-humid; temperate to warm winter.
  Northern atlantic Morocco (Hinterland of the littoral between Kenitra and Mohammadia).
- *Community of Crassula vaillantii* Molina, Tahiri, Agostinelli et al. 2009
  Thermomediterranean. Semi-arid, Sub-humid; temperate to warm winter.
  Northern atlantic Morocco (Hinterland of the littoral between Kenitra and Mohammadia).
- *Community of Cyperus michelianus* Molina, Tahiri, Agostinelli et al. 2009
  Thermomediterranean. About 100 m. Semi-arid, Sub-humid; temperate to warm winter. Loamy-clayey.
  Northern Atlantic Morocco (Maamora).
- *Community of Eryngium atlanticum* Molina, Tahiri, Agostinelli et al. 2009
  Thermomediterranean. Semi-arid, Sub-humid; temperate to warm winter.
  Northern atlantic Morocco (Hinterland of the littoral between Kenitra and Mohammadia).

M. S. Taleb, M. Fennane, *Vascular Plant Communities of Morocco*, Geobotany Studies, https://doi.org/10.1007/978-3-319-93704-5_5

Fig. 1 Aquatic plant communities (*Iris pseudacorus* dominant), region of Larache (ph. El Oualidi)

- *Community of Glyceria declinata* Molina, Tahiri, Agostinelli et al. 2009
  Thermomediterranean. Aroud 200 m.
  Northern Atlantic Morocco (Benslimane forest).
- *Community of Isoëtes velata* subsp. *adspersa* Molina, Tahiri, Agostinelli et al. 2009
  Thermomediterranean. Semi-arid, Sub-humid; temperate to warm winter.
  Northern atlantic Morocco (Hinterland of the littoral between Kenitra and Mohammadia).
- *Community of Oldenlandia capensis* Molina, Tahiri, Agostinelli et al. 2009
  Thermomediterranean. Semi-arid, Sub-humid; temperate to warm winter.
  Northern atlantic Morocco (Hinterland of the littoral between Kenitra and Mohammadia).
- *Community of Polypogon maritimus* Molina, Tahiri, Agostinelli et al. 2009
  Thermomediterranean. Semi-arid, Sub-humid; temperate to warm winter.
  Northern atlantic Morocco (Hinterland of the littoral between Kenitra and Mohammadia).
- *Association with Carex divisa* Nègre 1959
  Thermomediterranean. Arid; temperate winter. Unsalted muds.
  Middle atlantic Morocco (oued Nfiss).
- *Association with Cladium mariscus* Nègre 1959
  Thermomediterranean. Arid; temperate winter. Muds, non-current waters.
  Middle atlantic Morocco (Oued El-Arich).

Fig. 2 Aquatic plant communities (*Nymphaea alba* dominant), region of Larache (ph. Bennig)

- *Association with Festuca arundinacea* Nègre 1959
  Thermomediterranean. Arid; temperate winter. Unsalted muds.
  Middle atlantic Morocco (oued Ouerna in Bni-Moussa, Shallow water table all the year).
- *Association with Heliosciadium nodiflorum* Nègre 1959
  Thermomediterranean. Arid; temperate winter. Current water.
  Middle atlantic Morocco (Waters of streams, seguias and permanent wadis of the central Haouz and Tadla).
- *Association* with *Isoetes velata* and *Marsilea pubescens* Nègre 1959
  Thermomediterranean. Arid; temperate winter.
  Middle atlantic Morocco (Non-current water).
- *Association* with *Juncus martimus* and *Apium nodiflorum* Nègre 1959
  Thermomediterranean. Arid; temperate winter. Muds.
  Middle atlantic Morocco (current water, slightly salty at Ouljat Zaouia Sidi Mohamed Ben Sassi east of Marrakech).
- *Association* with *Ranunculus aquatilis* and *Limosella aquatica* Nègre 1959
  Thermomediterranean. Arid; temperate winter.
  Middle atlantic Morocco (dayas).
- *Association* with *Scirpus lacuster* and *Juncus maritimus* Nègre 1959
  Thermomediterranean. Arid; temperate winter. Muds.
  Middle atlantic Morocco (marshes covered with more than 30 cm of water).

Fig. 3 Aquatic plant communities (*Juncaceae* and *Poaceae* dominant), region of Mehdia (ph. Taleb)

Fig. 4 Aquatic plant communities (*Ranunculus aquatilis* dominant), Maamora suberia (ph. Grillas)

- *Association* with *Scirpus maritimus* Nègre 1959
  Thermomediterranean. Arid; temperate winter. Unsalted muds.
  Middle atlantic Morocco (Sedd el-Messjoun).
- *Association* with *Scirpus lacustris* and *Ranunculus aquaticus* Nègre 1961

Fig. 5 Aquatic plant communities (*Juncus* and *Phragmites* dominant), Sidi Bou Ghaba (ph. Taleb)

Fig. 6 Aquatic plant communities (*Juncaceae* dominant), region of Mehdia (ph. Taleb)

Mesomediterranean, Mountain mediterranean. Sub-humid, cold winter. Non-current waters.

Middle-Atlas (jbel Saa on the innermost margins of aguelmams with deep water at least 80 cm).

- *Association* with *Scutellaria orientalis* var. *demnatensis* Nègre 1961

  Mountain mediterranean. Sub-humid; cold winter. 2050 m. Volcanic ash more or less solidified by erosion.

  Central Middle-Atlas (north flank of the jbel Tamakouit).
- *Association* with *Tillaea vaillantii* and *Limosella aquatica* Nègre 1959

  Thermomediterranean. Arid; temperate winter.

  Middle atlantic Morocco (dayas).
- *Association* with *Trifolium subterraneum* Nègre 1959

  Thermomediterranean. Arid; temperate winter.

  Middle atlantic Morocco (dayas).
- *Association* with *Typha angustifolia* Nègre 1959

  Thermomediterranean. Arid; temperate winter. Muds and marshes.

  Middle atlantic Morocco (muds on the edge of wadis and marshes always covered with at least a few centimeters of water).
- *Ceratoneuro-Caricetum flavae* Nègre 1961

  Mountain mediterranean. Sub-humid; cold winter. Dolomite.

  Middle-Atlas (Tamarakouit: water courses).
- *Festucetum hystricis* Nègre 1961

  Mountain mediterranean. 2000 m. Sub-humid, Humid; cold winter. Marly-calcareous.

  Middle-Atlas (summit of jbel Saa).
- *Junco maritimae-Taraxacetum getulae* Nègre 1959

  Thermomediterranean. Arid; temperate winter.

  Middle atlantic Morocco (oued Nfiss).
- *Tetragonolobo siliquae-Taraxacetum getulae* Nègre 1959

  Thermomediterranean. Arid; temperate winter.

  Middle atlantic Morocco (oued Nfiss).
- *Myosuro-Convolvuletum* Nègre 1961

  Mesomediterranean, Mountain mediterranean. Sub-humid, Humid; cold to very cold winter. Basalt or clay from volcanic origin. Clayey.

  Middle-Atlas (dayas).
- *Ranunculo-Poetum trivialis* Nègre 1961

  Mountain mediterranean. Sub-humid; cold winter.

  Middle-Atlas (North flank of the jbel Tamarakouit; slightly stagnant water).
- *Groupement* with *Eleocharis palustris* Nègre 1961

  Mountain mediterranean. Sub-humid, Humid; cold winter.

  Middle-Atlas (jbel Saa: toward the outside of the aguelmames, the water level is a few cm above the ground and can even sink below the surface on the summer period).
- *Sedo-Juncetum pygmae* Quézel 1957

  Oromediterranean. 2700–3050 m. Semi-arid; very cold winter. Wet mud.

Central High-Atlas (Oukaimeden plateau on the dge of pozzines, pozzines of high Tessaout, north flank of Tizin'Ouano).

Phragmito-Magnocaricetea Klika in Klika et Novác 1941

Swampy, fenny and lacustrine habitats. Tall herbland helophyte vegetation dominated by perennial graminoids, sedges, bulrushes, reeds... Fresh and brackish waters.

Bolboschoenetalia maritimi *Hejny in Holub et al. 1967*

(*Scirpetalia compacti* Hejny in Holub et al. 1967 corr. Rivas-Martinez et al. 1980)

Scirpion maritimi Dahl and Hadac 1941

(*Scirpion compacti* Dahl and Hadac 1941 corr. Rivas-Martinez et al. 1980)

- *Crypsido aculeatae-Scirpetum compacti* Bendaanoun 1991
  Thermomediterranean. Semi-arid; warm winter. Clays.
  Rif (Smir, Martil).
- *Cynodo hirsutissimi-Scirpetum (maritimi) compacti* Bendaanoun 1991
  Thermomediterranean. Sub-humid; warm winter. Loamy sands, sandy loams, loams on sands.
  Northern atlantic Morocco (Bou Regreg).
- *Paspalo vaginati-Scirpetum (maritimi) compacti* Bendaanoun 1991
  Thermomediterranean. Sub-humid; temperate winter. Sandy mudds, mudds, loamy clays.
  Northern atlantic Morocco (Moulay Bou Selham).
- *Phragmito (australis) humilis-Scirpetum compacti* Bendaanoun 1991
  Thermomediterranean. Semi-arid; warm winter. Loamy clays.
  Rif (Smir).
- *Scirpetum (maritimi) compacti* Van Langendock 1931 corr. Bueno and F. Prieto in Bueno 1997
  Thermomediterranean. Semi-arid, Sub-humid; warm, temperate winter. Consolidated mudds, clayey loams.
  Middle atlantic Morocco (Oualidia) Northern atlantic Morocco (Tahadart, Loukkos, Bou Regreg) Rif (Smir).

Magnocaricetalia *Pignatti 1953*

Magnocaricion elatae Koch 1926

- *Cladietum marisci* (Allorge 1922) Zobrist 1935
  Thermomediterranean. Sub-humid; temperate winter. Soft mudds.
  Northern atlantic Morocco (Moulay Bou Selham, Loukkos, Sebou).

Nasturtio-Glycerietalia *Pignatti 1953*

Glycerio-Sparganion Braun-Blanquet and Sissingh in Boer 1942

- *Trifolio micheliani-Oenanthetum fistulosae* Deil 1997
  Thermomediterranean. Sub-humid; temperate and warm winter. Hydromorphic soils.
  Known in small areas in Tangerian peninsula (of Tangier) and in Spain (north west of Tarifa).

Other Groupements

- *Groupement with Helosciadium nodiflorum* Br.-Bl. and Maire 1924
  Thermomediterranean. Arid; temperate winter.
  Middle atlantic Morocco (Haouz).

Phragmitetalia *Koch 1926*

Phragmition communis Koch 1926

- *Phragmitetum altissimae* (Chapman 1960) em. Bendaanoun 1991
  Infra-, Thermomediterranean. Arid, Semi-arid, Sub-humid; temperate, warm winter. Clayey loams, mudds, sandy loams, loamy clays.
  Middle atlantic Morocco (Massa) Northern atlantic Morocco (Moulay Bou Selham) Rif (Smir) Mediterranean coast (Moulouya, Nador).
- *Phragmito (australis) altissimae-Scirpetum lacustris* Tüxen and Preising 1942
  Infra-, Thermomediterranean. Arid, Semi-arid, Sub-humid; temperate, warm winter. Mudds.
  Middle atlantic Morocco (Massa) Northern atlantic Morocco (Moulay Bou Selham), Rif (Negro, Smir).
- *Phragmito altissimae-Typhetum latifolio/angustifoliae* Bendaanoun 1991
  Infra-, Thermomediterranean. Arid, Semi-arid, Sub-humid; temperate, warm winter. Mudds.

Middle atlantic Morocco (Massa) Northern atlantic Morocco (Moulay Bou Selham), Rif (Negro, Smir).

- *Scirpetum lacustris* Bendaanoun 1991

  Infra-, Thermomediterranean. Arid, Semi-arid, Sub-humid; temperate, warm winter. Mudds.

  Middle atlantic Morocco (Massa) Northern atlantic Morocco (Moulay Bou Selham, Tahadart, Loukkos, Sebou, Bou Regreg) Rif (Negro, Smir, Martil) Mediterranean coast (Moulouya).
- *Scirpetum maritimi* (Braun-Blanquet 1931) Tüxen 1937

  Thermomediterranean. Semi-arid, Sub-humid; temperate, warm winter. Mudds, sandy mudds, clayey loams.

  Middle atlantic Morocco (Sidi Moussa) Northern atlantic Morocco (Loukkos, Moulay Bou Selham, Sebou, Bou Regreg).
- *Scirpetum tabernaemontani* Passagre 1964

  Thermomediterranean. Sub-humid; temperate winter. Consolidated mudds.

  Northern atlantic Morocco (Moulay Bou Selham, Loukkos, Sebou).
- *Scirpo lacustris-Phragmitetum altissimae* Braun-Blanquet 1931

  Thermomediterranean. Semi-arid; warm winter. Consolidated mudds.

  Rif (Smir, Negro, Martil).
- *Scirpo lacustris-Typhetum latifolio/angustifoliae* Bendaanoun 1991

  Infra-, Thermomediterranean. Arid, Semi-arid, Sub-humid; temperate, warm winter. Consolidated mudds.

  Middle atlantic Morocco (Massa), Northern atlantic Morocco (Bou Regreg, Moulay Bou Selham), Rif (Smir, Negro), Mediterranean coast (Moulouya).
- *Tamarico (gallicae) lagunae-Phragmitetum altissimae* Bendaanoun 1991

  Thermomediterranean. Semi-arid; warm winter. Loamy clays.

  Rif (Smir, Negro).
- *Typhetum latifolio-angustifoliae* Bendaanoun 1991

  Infra-, Thermomediterranean. Arid, Semi-arid, Sub-humid; temperate, warm winter. Soft mudds.

  Middle atlantic Morocco (Massa) Northern atlantic Morocco (Moulay Bou Selham, Sebou, Bou Regreg) Rif (Smir, Negro) Mediterranean coast (Moulouya).
- *Typho angustifoliae-Scirpetum lacustris* (Braun-Blanquet and Bolos 1958) Rivas-Martinez 1966

  Infra-, Thermomediterranean. Arid, Semi-arid, Sub-humid; temperate, warm winter. Mudds.

  Middle atlantic Morocco (Massa) Northern atlantic Morocco (Bou Regreg, Loukkos) Rif (Smir) Mediterranean coast (Moulouya).

Other Groupements

- *Community of Apium nodiflorum subsp. mairei* Molina, Tahiri, Agostinelli et al. 2009

  Thermomediterranean. Around 100 m. Sub-humid, Semi-arid; temperate to warm winter.

  Northern atlantic Morocco (Hinterland of the littoral between Kenitra and Mohammadia).

- *Community of Eleocharis palustris* Molina, Tahiri, Agostinelli et al. 2009
  Thermomediterranean. 40–120 m. Sub-humid, Semi-arid; temperate to warm winter. Clayey.
  Northern atlantic Morocco (Lagoons and ponds of the Maamora; atlantic coast of the Tangerian peninsula).
- *Community of Damasonium bourgaei* Molina, Tahiri, Agostinelli et al. 2009
  Thermomediterranean. 200 m. Semi-arid; temperate winter. Clayey.
  Northern atlantic Morocco (Ponds of the Benslimane forest).
- *Association with Ormenis nobilis var. discoidea* Br.-Bl. and Maire 1924
  Mesomediterranean, 1200–1250 m. Subhumid; cold winter. Basalte.
  Central Middle-Atlas: wet depression where water stays in winter for a short time.
- *Association with Pulicaria arabica* Br.-Bl. and Maire 1924
  Mesomediterranean, 1200–1250 m. Subhumid; cold winter. Basalte.
  Central Middle-Atlas: wet depression where water stays in winter for a long time.
- *Association with Senecio giganteus* Br.-Bl. and Maire 1924
  Mesomediterranean, 1200–1300 m. Subhumid; cold winter.
  Central Middle-Atlas (edges of watercourses aroud Tioumliline).

Potamogetonetea Klika in Klika and Novák 1941

Rooted herbs floating or submerged in water fresh or brackish. Stagnant water and/or slowly flowing shallow streams.

- *Community of Potamogeton nodosus* Molina, Tahiri, Agostinelli et al. 2009
  Thermomediterranean. 100 m. Sub-humid, Semi-arid; temperate, warm winter.
  Northern atlantic Morocco (Hinterland of the littoral between Kenitra and Mohammadia).
- *Community of Ranunculus saniculifolius* Molina, Tahiri, Agostinelli et al. 2009
  Thermomediterranean. 80–110 m. Semi-arid; temperate winter.
  Northern atlantic Morocco (Maamora, bled Dendoun, Sidi Amira).

Terrestrial (Inland and Coastal) Halo-Gypsophile Vegetation

Atriplicetea halimi Nègre 1959

Caracteristic species [Nègre 1959: 255]: *Atriplex halimus, Limonium thouini, Salsola vermiculata var. flavescens, Vella annua.*

Atriplicetalia halimi *Nègre 1959*

Caracteristic species (Nègre 1959: 255): *Ammi majus var. tenue, Anacyclus valentinus, Bubonium odorum, Diplotaxis assurgens, Phalaris brachystachys, Vicia benghalensis var. atropurpurea, Volutaria crupinoides.*

- *Association with Atriplex halimus and Salsola vermiculata* Br.-Bl. and Maire 1924.
  Thermomediterranean. Arid; temperate winter.
  Middle atlantic Morocco (Haouz) [Braun-Blanquet and Maire 1924: 28].

Salsoletalia vermiculatae *Nègre 1959*

Caracteristic species [Nègre 1959: 255]: *Aizoon hispanicum, Bupleurum semicompositum, Frankenia corymbosa.*

- *Salsolo vermiculatae-Trigonelletum anguinae* Nègre 1959
  Thermomediterranean. Arid; temperate winter. Limestone, loamy-clayey.
  Middle atlantic Morocco (bled El-Harousia north of oued Tensift).

M. S. Taleb, M. Fennane, *Vascular Plant Communities of Morocco*, Geobotany Studies, https://doi.org/10.1007/978-3-319-93704-5_6

Nerio-Tamaricetea Br.-Bl. and O. de Bolos 1958

Riparian and lacustrine woodlands, dominated by *Tamarix spp, Nerium oleander, Phragmites*. . ., without running water during long periods of the year or temporarily inundated by fresh, brackish or saline water.

Tamaricetalia africanae *Br.-Bl. and O. de Bolos 1958*

Tamaricion africanae Br.-Bl. and O. de Bolos 1958

- *Phragmito altissimae-Tamaricetum africanae* Bendaanoun 1991
  Infra-, Thermomediterranean. Arid, Semi-arid; temperate and warm winter. Clays, loams.
  Middle atlantic Morocco (Massa) Rif (Negro, Smir, Martil) Mediterranean coast (Moulouya).
- *Sarcocornio alpini-Tamaricetum africano/lagunae* Bendaanoun 1991
  Thermomediterranean. Semi-arid; temperate, warm winter. Clays, loams.
  Northern atlantic Morocco (oued Mellah) Rif (Martil).
- *Scirpo compacti-Tamaricetum africanae* Bendaanoun 1991
  Infra-, Thermomediterranean. Arid, Semi-arid; temperate and warm winter. Clays, loams.
  Middle atlantic Morocco (Massa) Northern atlantic Morocco (oued Mellah) Rif (Smir) Mediterranean coast (Moulouya).
- *Spartino (junceae) patentis-Tamaricetum africanae* Bendaanoun 1991
  Thermomediterranean. Semi-arid; warm winter. Loamy clays on sands.
  Rif (Smir, Negro).
- *Inulo crithmoidis-Tamaricetum boveano/africanae* Izco et al. 1984 em. Bendaanoun 1991
  Thermomediterranean. Semi-arid; warm winter. Loams on sands.
  Mediterranean coast (Moulouya).
- *Sarcocornio alpini-Tamaricetum boveano/africanae* Bendaanoun 1991
  Thermomediterranean. Semi-arid; warm, temperate winter. Clayey loams, sandy loams.
  Mediterranean coast (Moulouya).
- *Arthrocnemo glauci-Tamaricetum boveano/malenconianae* Bendaanoun 1991
  Inframediterranean. Arid; warm winter. Loams on sands.
  Middle atlantic Morocco (Souss).
- *Sarcocornio fruticosae-Tamaricetum boveano/malenconianae* Bendaanoun 1991
  Inframediterranean. Arid; warm winter. Loams on sands, sandy loams.
  Middle atlantic Morocco (Souss).

Pegano harmalae-Salsoletea vermiculatae Br.-Bl. and O. de Bolos 1958

Halo-nitrophilous dwarf shrub communities in arid and Saharan coastal and inland habitats.

Salsolo-Nitrarietalia *Quézel 1965*

Limoniastro-Nitrarion Braun-Blanquet 1949

- *Lycio intricati-Rhusetum albidi* Gehu and Biondi 1996
  Thermomediterranean. Semi-arid; warm winter. Limestone slabs littoral more or less lapiazed, and more or less covered with sand.
  Middle atlantic Morocco (Entre Essaouira et Safi).
  Curious association of thorny brush, usually prostrate on the ground, high a few decimeters.
- *Lycio intricati-Retametum monospermae* Gehu and Biondi 1996
  Inframediterranean, Thermomediterranean. Arid, Semi-arid; warm winter. Sandy sandstone.
  Middle atlantic Morocco (Souss-Massa estuary, South Agadir, region of Tamri towards the North).
  Formation of Retama monosperma very marked physiognomically in the Landscape. It develops on the old sandy terraces of red sand, encrusted and set back from the current coastal systems.

Limoniastro-Zygophyllion Quézel 1965

- *Limoniastro-Nitrarietum retusae* Quézel 1965
  Saharan bioclimat; temperate winter. Salted substrates.
  Wadi beds where the water table is deep enough (Oued Draa, Oued Noun, Oued Hafra, Seguia el-Hamra).

Other Groupements

- *Artemisio reptantis-Salsoletum longifoliae* Bendaanoun 1991
  Thermomediterranean. Semi-arid; warm, temperate winter.
  Middle atlantic Morocco (Safi).
- *Frankenio (corymbosae) gracilis-Suaedetum longifoliae* Bendaanoun 1991
  Thermomediterranean. Arid, Semi-arid; temperate winter. Clayey loams.
  Middle atlantic Morocco (Zima).
- *Frankenio (corymbosae) gracilis-Atriplicetum halimi* Bendaanoun 1991

Thermomediterranean. Arid, Semi-arid; temperate winter. Sandy loams, loams.

Middle atlantic Morocco (Zima).

- *Lycio intricati-Salsoletum longifoliae* Bendaanoun 1991

Thermomediterranean. Semi-arid; warm, temperate winter. Colluvium on sandstone.

Middle atlantic Morocco (Safi).

- *Plantago litardieri-Atriplecetum malvanae* Bendaanoun 1991

Mesomediterranean. Semi-arid; cold winter. Loams.

Eastern plateaux (Missour).

- *Spergulario fimbriatae-Frankenietum velutinae* Bendaanoun 1991

Thermomediterranean. Semi-arid; warm, temperate winter. Loamy sands on sandstone, sand on sandstone.

Middle atlantic Morocco (Casablanca) Northern atlantic Morocco (Rabat, Témara, Skhirat).

- *Spergulario fimbriatae-Suaedetum longifoliae* Bendaanoun 1991

Thermomediterranean. Semi-arid; temperate winter. Loamy sands, sands.

Middle atlantic Morocco (Oualidia, Sidi Moussa).

Saginetea maritimae Westhoff, Van Leeuwen, and Adriani 1962

(*Mesembryanthemetea nodiflori* Nègre (1956) 1959)
(*Frankenietea pulverulentae* Rivas-Mart. in Rivas-Martinez and Costa 1976)
Coastal and inland ephemeral vegetation in disturbed saline habitats and saline badlands.

- *Aeluropetum litoralis* Nègre (1956) 1959

Thermomediterranean. Arid; temperate winter. Salty soils (chlorures and sulfates), generally immersed for a short time of the year.

Middle atlantic Morocco (Sedd El-Messjoun).

Quite heterogeneous association, five facies distinguished:

– *facies with Aeluropus litoralis*
– *facies with Melilotus indica*
– *facis with Coronopus procumbens*
– *facies with Beta macrocarpa*
– *facies with Mesembryanthemum nodiflorum*

- *Anabasetum aphyllae* Nègre 1959

  Thermomediterranean. Arid; temperate, cool winter. Calcareous marls, alluviums.

  Middle atlantic Morocco (Haouz southern Chichaoua (Nègre 1959:165)) Eastern plateaux (region of Guercif, Taourirt…(Braun-Blanquet and Maire 1924:22)).

- *Frankenio corymbosae-Koelpinietum linearis* Nègre (1956) 1959

  Thermomediterranean. Arid; temperate winter. Sandy soils, salty (chlorures and sulfates), never saturated with water.

  Middle atlantic Morocco (Sedd El-Messjoun).

  Subassociations:

  – *diplotaxisetosum ollivieri*
  – *mesembryanthemetosum nodiflori*
  – *facies with Anabasis aphylla*

- *Frankenio (corymbosae) phosphaticae-Limonietum thouini* Nègre 1959

  Thermomediterranean. Arid; temperate winter. Alluviums.

  Middle atlantic Morocco (Haouz).

- *Mesembryanthemo-Spergularietum diandrae* Nègre (1956) 1959

  Thermomediterranean. Arid; temperate winter. Stony substrates (small pebble mounds); salty soils (chlorures and sulfates), never saturated with water.

  Middle atlantic Morocco (Haouz of Marrakech: Sedd El-Messjoun, Sidi Zouine…).

Salicornietea fruticosae Br.-Bl. and Tüxen in A. Bolos y Vayreda 1950

(*Sarcocornietea fruticosae* Br.-Bl. and Tx. ex A. Bolos y Vayreda 1950)

Perennial salt-marsh herblands and shrub ; temporarily inundated succulent chenopodiaceous dwarf shrubs. (Figs. 1, 2, 3, 4 and 5).

Limoniastrietalia guoynoniani *Guinochet 1951*

Limoniastrion monopetali Pignatti 1953

- *Artemisio (coerulescentis) gallicae-Limoniastretum monopetali* Baudière et al. 1976

  Thermomediterranean. Sub-humid; temperate winter. Loamy clays.

  Northern atlantic Morocco (Loukkos).

Fig. 1 Halophytes community (*Salicornietea*), Sahara (ph. Ibn Tattou)

Fig. 2 Halophytes community (*Salicornietea*), region of Larache (ph. Bennig)

Fig. 3 Halophytes community (*Mesembryanthemum nodiflorum* dominant)

- *Arthrocnemo glauci-Limoniastretum monopetali* Bendaanoun 1991
  Thermomediterranean. Sub-humid; temperate winter. Loamy sands.
  Northern atlantic Morocco (Bou Regreg).
- *Limonio ferulacei-Limoniastretum monopetali* Bendaanoun 1991
  Thermomediterranean. Sub-humid; temperate winter. Sands.
  Northern atlantic Morocco (Tahadart).
- *Limonio lychnidifolii-Limoniastretum monopetali* Braun-Blanquet 1952
  Thermomediterranean. Sub-humid; temperate winter. Sandy loams on sands, sandy clays, consolidated vases, sandy vases, clays.
  Northern atlantic Morocco (Bou Regreg, Loukkos).
- *Limonio vulgaris-Limoniastretum monopetali* Bendaanoun 1991
  Thermomediterranean. Sub-humid; temperate winter. Sands, loamy clays, clayey sands.
  Northern atlantic Morocco (Tahadart).
- *Suaedo brevifoliae-Limoniastretum monopetali* Bendaanoun 1991
  Thermomediterranean. Sub-humid; temperate winter. Loams, clayey loams.
  Northern atlantic Morocco (Bou Regreg).
- *Triglochino barrelieri-Limoniastretum monopetali* Bendaanoun 1991
  Thermomediterranean. Sub-humid; temperate winter. Clays, loamy clays.
  Northern atlantic Morocco (Bou Regreg).

Fig. 4 Halophytes community (*Arthrocnemum*, *Mesembryanthemum* and *Limonium* dominant), Sahara (ph. Ibn Tattou)

Saliconietalia fruticosae *Br.-Bl. 1933*

(*Sarcocornietalia fruticosae* Br. Bl. 1933)

Arthrocnemion glauci Rivas-Martinez and Costa 1984

- *Cresso criticae-Sarcocornietum alpini* Bendaanoun 1991
  Thermomediterranean. Arid, Semi-arid, Sub-humid; temperate, warm winter. Sandy loams, loams, clayey loams on sands.
  Middle atlantic Morocco (Zima) Northern atlantic Morocco (Tahadart, Gharifa, Loukkos, Bou Regreg) Mediterranean coast (Moulouya).
- *Frankenio (corymbosae) genuinae-Arthrocnemetum glauci* Bendaanoun 1991
  Thermomediterranean. Semi-arid; temperate, warm winter. Sands.
  Mediterranean coast (Moulouya, Nador).
- *Frankenio (corymbosae) laxae-Arthrocnemetum glauci* Bendaanoun 1991
  Infra-, Thermomediterranean. Arid, Semi-arid; warm, temperate winter. Sandy loams.
  Middle atlantic Morocco (Souss).
- *Frankenio (corymbosae) leucanthae-Arthrocnemetum glauci* Bendaanoun 1991

Fig. 5 Halophytes community (*Chenopodiaceae* dominant), region of Khnifiss (ph. Ibn Tattou)

Infra-, Thermomediterranean. Arid, Semi-arid; warm, temperate winter. Loamy sands, sands.

Middle atlantic Morocco (Souss, Massa).

- *Frankenio laevis-Arthrocnemetum glauci* Bendaanoun 1991

  Thermomediterranean. Semi-arid; warm, temperate winter. Loams, fine sands.

  Northern atlantic Morocco (Gharifa) Rif (Martil).

- *Frankenio pallidae-Arthrocnemetum glauci* Bendaanoun 1991

  Thermomediterranean. Semi-arid; warm, temperate winter. Loamy clays.

  Mediterranean coast (Moulouya).

- *Frankenio pulverulentae-Arthrocnemetum glauci* Bendaanoun 1991

  Infra-, Thermomediterranean. Arid, Semi-arid; warm, temperate winter. Loams.

  Middle atlantic Morocco (Massa) Northern atlantic Morocco (Bou Regreg) Mediterranean coast (Moulouya).

- *Hordeo maritimae-Arthrocnemetum glauci* Bendaanoun 1991

  Thermomediterranean. Saharan, Arid, Semi-arid, Sub-humid; temperate, warm winter. Clays, sandy loams, clayey loams, clays.

  Saharan Morocco (Draa) Middle atlantic Morocco (Oum R'bia, Oualidia, Mellah) Northern atlantic Morocco (Tahadart, Gharifa, Loukkos, Bou Regreg) Rif (Martil).

- *Hordeo maritimae-Sarcocornietum alpini* Bendaanoun 1991

Thermomediterranean. Semi-arid, Sub-humid; warm, temperate winter. Clayey loams.

Northern atlantic Morocco (Tahadart, Gharifa, Loukkos, Bou Regreg).

- *Limonio (cymuliferi) sebkhari-Arthrocnemetum glauci* Bendaanoun 1991

Thermomediterranean. Semi-arid; warm, temperate winter. Sands, loamy sands.

Mediterranean coast (Moulouya, Nador).

- *Limonio asparagoidis-Arthrocnemetum glauci* Bendaanoun 1991

Thermomediterranean. Semi-arid; warm, temperate winter. Loams, clayey loams.

Mediterranean coast (Nador).

- *Limonio asparagoidis-Sarcocornietum alpini* Bendaanoun 1991

Thermomediterranean. Semi-arid; warm, temperate winter. Loams, clayey loams.

Mediterranean coast (Nador).

- *Limonio ferulacei-Arthrocnemetum glauci* Bendaanoun 1991

Thermomediterranean. Semi-arid, Sub-humid; warm, temperate winter. Sands, loamy sands, sandy loams, clayey loams on sands.

Middle atlantic Morocco (Oualidia, Oum R'bia) Northern atlantic Morocco (Tahadart, Gharifa, Mellah, Loukkos, Bou Regreg).

- *Limonio vulgaris-Arthrocnemetum glauci* Bendaanoun 1991

Thermomediterranean. Sub-humid; temperate winter. Sands, sandy clays.

Northern atlantic Morocco (Tahadart, Gharifa).

- *Mesembryanthemo nodiflori-Arthrocnemetum glauci* Bendaanoun 1991

Infra-, Thermomediterranean. Arid, Semi-arid; temperate winter. Sands, sandy loams.

Middle atlantic Morocco (Souss, Massa, Oualidia, Zima) Mediterranean coast (Moulouya).

- *Obiono portulacoidis-Arthrocnemetum glauci* Bendaanoun 1991

Thermomediterranean. Semi-arid, Sub-humid; temperate, warm winter. Loamy sands, clays, loams.

Middle atlantic Morocco (Oum R'bia, Sidi Moussa, Oualidia, Mellah) Northern atlantic Morocco (Tahadart, Gharifa, Loukkos, Bou Regreg) Rif (Martil) LM (Rhiss, Nekkor, Nador).

- *Phragmito altissimae-Arthrocnemetum glauci* Bendaanoun 1991

Infra-, Thermomediterranean. Arid, Semi-arid; warm, temperate winter. Clays on sands, clayey loams on sands.

Middle atlantic Morocco (Souss).

- *Phragmito humilis-Sarcocornietum alpini* Bendaanoun 1991

Thermomediterranean. Semi-arid, Sub-humid; warm, temperate winter. Clays on sands, loamy clays on sands, sandy loams, clayey loams on sands.

Northern atlantic Morocco (Moulay Bou Selham) Mediterranean coast (Moulouya).

- *Salicornio ramosissimae-Arthrocnemetum glauci* Bendaanoun 1991

Infra-, Thermomediterranean. Arid, Semi-arid, Sub-humid; warm, temperate winter. Clays, loamy clays, loamy sands, sandy loams.

Middle atlantic Morocco (Massa, Oum R'bia, Oualidia, Sidi Moussa, Mellah) Northern atlantic Morocco (N'fifikh, Tahadart, Gharifa, Loukkos) Rif (Martil) Mediterranean coast (Moulouya).

- *Sarcocornio alpini-Arthrocnemetum glauci* Bendaanoun 1991

  Infra-, Thermomediterranean. Arid, Semi-arid, Sub-humid; warm, temperate winter. Loamy clays, sandy clays, clayey loams.

  Middle atlantic Morocco (Souss, Oualidia, Sidi Moussa) Northern atlantic Morocco (Tahadart, Gharifa, Loukkos, Bou Regreg) Rif (Martil) Mediterranean coast (Moulouya, Nador).
- *Sphenopo divaricati-Arthrocnemum glauci* Braun-Blanquet (1928) 1933

  Infra-, Thermomediterranean. Arid, Semi-arid, Sub-humid; warm, temperate winter. Clayey loams, loams.

  Middle atlantic Morocco (Souss, Oum R'bia, Oualidia, Sidi Moussa) Northern atlantic Morocco (Tahadart, Gharifa, Loukkos, Bou Regreg)
- *Tamarico boveanae-Sarcocornietum alpini* Bendaanoun 1991

  Infra-, Thermomediterranean. Arid, Semi-arid; warm, temperate winter. Sandy loams.

  Middle atlantic Morocco (Souss) Mediterranean coast (Moulouya).
- *Tamarico boveano/malenconiane-Arthrocnemum glauci* Bendaanoun 1991

  Infra-, Thermomediterranean. Arid, Semi-arid; warm, temperate winter. Sandy loams.

  Middle atlantic Morocco (Souss).
- *Tamarico lagunae-Sarcocornietum alpini* Bendaanoun 1991

  Thermomediterranean. Semi-arid; warm, temperate winter. Loamy sands.

  Rif (Martil).
- *Zygophylletum fontanesii* Bendaanoun 1991

  Thermomediterranean. Semi-arid, Sub-humid; warm, temperate winter. Sandstones, sands.

  Middle atlantic Morocco (Sidi Moussa, Oualidia) Northern atlantic Morocco (Skhirat, Témara, Rabat).
- *Zygophyllo fontanesii-Arthrocnemum glauci* Fernandez and Santos 1983

  Thermomediterranean. Semi-arid; temperate winter. Coarse sands.

  Middle atlantic Morocco (Sidi Abed, Sidi Moussa, Oualidia).
- *Zygophyllo fontanesii-Limonietum ferulacei* Bendaanoun 1991

  Thermomediterranean. Semi-arid; temperate winter. Sands.

  Middle atlantic Morocco (Sidi Abed, Sidi Moussa, Oualidia).

Salicornion fruticosae Br.-Bl. 1933

(*Sarcocornion fruticosae* Br.-Bl. 1933)

- *Artemisio (coerulescentis) gallicae-Sarcocornietum fruticosae* Bendaanoun 1991

Thermomediterranean. Sub-humid; temperate winter. More or less loamy-clayey.

Northern atlantic Morocco (Loukkos).

- *Astero (tripolii) pannonici-Inuletum crithmoidis* Bendaanoun 1991
  Thermomediterranean. Sub-humid; temperate winter. Loamy clayey.
  Northern atlantic Morocco (Gharifa).
- *Astero (tripolii) pannonici-Sarcocornietum fruticosae* Bendaanoun 1991
  Thermomediterranean. Sub-humid; temperate winter. Sandy vase, consolidated vase.
  Northern atlantic Morocco (Tahadart, Gharifa).
- *Bostrychio scorpioidis-Obionetum portulacoides* (Corill.) Tüxen (1953) 1963
  Thermomediterranean. Semi-arid, Sub-humid; temperate, warm winter. Sandy vase, sandy loam, loamy clay, loamy sand.
  Middle atlantic Morocco (Oualidia, Sidi Moussa) Northern atlantic Morocco (Tahadart, Gharifa, Bou Regreg, Moulay Bou Selham, Loukkos).
- *Bostrychio scorpioidis-Sarcocornietum perennis* Bendaanoun 1991
  Thermomediterranean. Semi-arid, Sub-humid; temperate, warm winter. Sandy vase.
  Middle atlantic Morocco (Oualidia) Northern atlantic Morocco (Tahadart, Gharifa, Bou Regreg).
- *Cistancho phelipae-Sarcocornietum fruticosae* Gehu (1973) 1976
  Thermomediterranean. Semi-arid, Sub-humid; temperate, warm winter. Consolidated muddy-sand substrate.
  Northern atlantic Morocco (Bou Regreg, Loukkos) [Bendaanoun 1991:312].
  Subassociations:

  – *limoniastretosum monopetali*
  – *Inuletosum crithmoides*

- *Frankenio (corymbosae) genuinae-Sarcocornietum fruticosae* Bendaanoun 1991
  Thermomediterranean. Semi-arid; temperate, warm winter. Loam, sandy loam.
  Mediterranean coast (Moulouya, Nador).
- *Frankenio (corymbosae) thymoidis-Sarcocornietum fruticosae* Bendaanoun 1991
  Thermomediterranean. Semi-arid; temperate, warm winter. Sand, loamy sand.
  Mediterranean coast (Moulouya, Nador).
- *Frankenio laevis-Sarcocornietum perennis* Bendaanoun 1991
  Thermomediterranean. Semi-arid, Sub-humid; temperate, warm winter. Clayey loams.
  Northern atlantic Morocco (Bou Regreg) Rif (Martil).
- *Fuco axillaris-Sarcocornietum perennis* Bendaanoun 1991
  Thermomediterranean. Semi-arid; temperate winter. Muddy sands.
  Middel atlantic Morocco (Oualidia, Sidi Moussa).
- *Inulo chritmoidis-Sarcocornietum fruticosae* Br.-Bl. 1931
  Thermomediterranean. Semi-arid, Sub-humid; temperate, warm winter.

Northern atlantic Morocco (littoral from Walidia to Tahadart) Rif (Martil, ...) Mediterranean coast (Nekkor, Ghiss, Nador, Moulouya...).

Heteregenous association, little known.

Subassociations (Bendaanoun 1991):

- *suaedetosum (maritimae) perennantis*
- *juncetosum maritimae*
- *juncetosum subulati*
- *limonietosum ferulacei*
- *phragmitetosum altissimae*
- *tamaricetosum (gallicae) lagunae*

• *Junco subalati-Sarcocornietum fruticosae* Bendaanoun 1991
Thermomediterranean. Semi-arid, Sub-humid; temperate, warm winter. Loamy clays.
Northern atlantic Morocco (Moulay Bou Selham, Bou Regreg) Rif (Smir).

• *Limonio (cymuliferi) sebkhari-Sarcocornietum fruticosae* Bendaanoun 1991
Thermomediterranean. Semi-arid; temperate, warm winter. Coarse sands.
Mediterranean coast (Nador, Moulouya).

• *Limonio (delicatuli) leptostachydis-Sarcocornietum fruticosae* Bendaanoun 1991
Thermomediterranean. Semi-arid; temperate, warm winter. Clayey sands.
Mediterranean coast (Moulouya).

• *Limonio asparagoidis-Sarcocornietum fruticosae* Bendaanoun 1991
Thermomediterranean. Semi-arid; temperate, warm winter. Clayey loams.
Mediterranean coast (Nador).

• *Limonio ferulacei-Obionetum portulacoidis* Bendaanoun 1991
Thermomediterranean. Semi-arid; temperate winter. Sands, loams, sandy loams, clayey loams.
Middle atlantic Morocco (Sidi Abed, Sidi Moussa, Oualidia).

• *Limonio gummiferae-Inuletum chrimoidis* Bendaanoun 1991
Thermomediterranean. Semi-arid; temperate, warm winter. Sands, loamy sands.
Mediterranean coast (Nador, Moulouya).

• *Limonio vulgaris-Inuletum chrimoidis* Bendaanoun 1991
Thermomediterranean. Semi-arid, Sub-humid; temperate, warm winter. Clayey loams, sandy loams, loams, consolided vase.
Middle atlantic Morocco (Oualidia) Northern atlantic Morocco (Tahadart, Gharifa, Loukkos).

• *Obionetum portulacoidis* Des Abb. and Corill. 1949
Thermomediterranean. Semi-arid, Sub-humid; temperate, warm winter. Sandy loams, loamy sands.
Middle atlantic Morocco (Oualidia) Northern atlantic Morocco (Tahadart, Loukkos).

• *Obiono portulacoidis-Puccinellietum maritimae* Gehu 1976
Thermomediterranean. Sub-humid; temperate, warm winter. Loamy sands, sandy loams.

Northern atlantic Morocco (Bou Regreg).

- *Obiono portulacoidis-Sarcocornietum fruticosae* Bendaanoun 1991
  Thermomediterranean. Semi-arid, Sub-humid; temperate, warm winter. Clayey loams, consolidated vase, sandy vase.
  Middle atlantic Morocco (Oualidia) Northern atlantic Morocco (Tahadart, Gharifa, Bou Regreg, Loukkos).
- *Obiono portulacoidis-Suaedetum perennantis* Bendaanoun 1991
  Thermomediterranean. Semi-arid, Sub-humid; temperate, warm winter. Sandy vase, consolidated vase, muddy sands.
  Middle atlantic Morocco (Oualidia, Sidi Moussa) Northern atlantic Morocco (Tahadart, Gharifa, Loukkos, Bou Regreg, Mellah).
- *Phragmito altissimae-Sarcocornietum fruticosae* Bendaanoun 1991
  Infra-, Thermomediterranean. Arid, Semi-arid; temperate, warm winter. Sandy loams, clayey loams on sands, loamy clays on sands.
  Middle atlantic Morocco (Souss) Rif (Martil) Mediterranean coast (Nador, Nekkor).
- *Puccinellietum maritimae* (Warming 1980) Br.-Bl. and De Leeuw 1936
  Thermomediterranean. Sub-humid; warm, temperate winter.
  Northern atlantic Morocco (Bou Regreg).
- *Puccinellio fontqueri-Inuletum chritmoidis* Bendaanoun 1991
  Thermomediterranean. Sub-humid; temperate winter. Clayey loams on sands, sandy loams on sands, loamy sands.
  Northern atlantic Morocco (Tahadart, Gharifa, Loukkos).
- *Puccinellio fontqueri-Sarcocornietum fruticosae* Bendaanoun 1991
  Thermomediterranean. Sub-humid; temperate winter. Sandy vases, consolidated vases, clayey loams on sands.
  Northern atlantic Morocco (Tahadart, Gharifa, Loukkos).
- *Puccinellio fontqueri-Sarcocornietum perennis* Bendaanoun 1991
  Thermomediterranean. Sub-humid; temperate winter. Muddy sands.
  Northern atlantic Morocco (Tahadart, Loukkos).
- *Puccinellio fontqueri-Suaedetum perennantis* Bendaanoun 1991
  Thermomediterranean. Sub-humid; temperate winter. Sandy loams, sandy mudds, consolidated sandy mudds, consolidated mudds.
  Northern atlantic Morocco (Tahadart, Loukkos, Gharifa).
- *Puccinellio maritimae-Sarcocornietum fruticosae* (Arenes) Gehu (1933) 1976
  Thermomediterranean. Sub-humid; temperate, warm winter. Muddy sand, sandy vases, consolidated vases.
  Northern atlantic Morocco (Bou Regreg).
- *Puccinellio maritimae-Sarcocornietum perennis* (Arenes) Gehu (1933) 1976
  Thermomediterranean. Sub-humid; temperate, warm winter. Sandy vases.
  Northern atlantic Morocco (Bou Regreg).
- *Puccinellio maritimae-Suaedetum perennantis* Bendaanoun 1991
  Thermomediterranean. Semi-arid, Sub-humid; temperate, warm winter. Consolidated sandy vases, consolidated vases on sands.
  Northern atlantic Morocco (Bou Regreg, Mellah).

- *Salicornio fruticosae-Spergularietum sauvagei* Nègre (1956) 1959
  Thermomediterranean. Arid; temperate winter. Soils sandy, salty (mainly chlorures), saturated with water during a long period of the year.
  Middle atlantic Morocco (Haouz of Marrakech: Sedd El-Messjoun, Sidi Zouine…).
- *Sarcocornio alpini-Sarcocornietum fruticosae* Bendaanoun 1991
  Infra-, Thermomediterranean. Arid, Semi-arid, Sub-humid; temperate, warm winter. Sandy loams, clayey loams.
  Middle atlantic Morocco (Souss, Tensift) Northern atlantic Morocco (Mellah, Bou Regreg).
- *Spergulario embergeri-Sarcocornietum perennis* Bendaanoun 1991
  Thermomediterranean. Semi-arid; temperate winter. Muddy sands, sands.
  Middle atlantic Morocco (Oualidia, Sidi Moussa).
- *Suaedetum perennantis* Bendaanoun 1991
  Thermomediterranean. Semi-arid, Sub-humid; temperate, warm winter. Sandy vases, sandy loams.
  Middle atlantic Morocco (Oualidia) Northern atlantic Morocco (Tahadart, Loukkos, Gharifa, Bou Regreg).
- *Suaedo perennantis-Sarcocornietum fruticosae* Bendaanoun 1991
  Thermomediterranean. Sub-humid; temperate, warm winter. Sandy vases, consolidated vases, consolidated vases on sands.
  Northern atlantic Morocco (Tahadart, Gharifa, Loukkos, Bou Regreg).
- *Triglochino maritimae-Sarcocornietum perennis* Bendaanoun 1991
  Thermomediterranean. Sub-humid; temperate winter. Sandy vases.
  Northern atlantic Morocco (Moulay Bou Selham).
- *Zygophyllo fontanesii-Inuletum crithmoidis* Bendaanoun 1991
  Thermomediterranean. Semi-arid; temperate winter. Sand on sandstone.
  Middle atlantic Morocco (Sidi Abed, Oualidia, Sidi Moussa).
- *Zygophyllo fontanesii-Obionetum portulacoidis* Bendaanoun 1991
  Thermomediterranean. Semi-arid; temperate winter. Sandy loams, sands, loamy sands.
  Middle atlantic Morocco (Sidi Abed, Sidi Moussa, Oualidia).
- *Zygophyllo fontanesii-Sarcocornietum fruticosae* Bendaanoun 1991
  Thermomediterranean. Semi-arid; temperate winter. Coarse sands.
  Middle atlantic Morocco (Sidi Abed, Sidi Moussa, Oualidia).

Suaedion brevifoliae Br.-Bl. and O. de Bolos 1958

(*Suaedion braun-blanquetii* Br.-Bl. and O. de Bolos 1958 corr. Rivas-Martinez et al. 1991a, b)

- *Frankenio (corymbosae) laxae-Suaedetum brevifoliae* Bendaanoun 1991
  Infra-, Thermomediterranean. Arid, Semi-arid; temperate, warm winter. Sandy loams, loamy sands.

Middle atlantic Morocco (Tensift, Souss, Massa).

- *Frankenio (corymbosae) leucanthae-Suaedetum brevifoliae* Bendaanoun 1991
  Infra-, Thermomediterranean. Arid; temperate, warm winter. Sands, sandy loams.
  Middle atlantic Morocco (Souss, Massa).
- *Frankenio (corymbosae) thymoidis-Suaedetum brevifoliae* Bendaanoun 1991
  Thermomediterranean. Semi-arid; temperate, warm winter. Clayey loams.
  North eastern Morocco (Moulouya).
- *Frankenio intermediae-Suaedetum brevifoliae* Bendaanoun 1991
  Thermomediterranean. Sub-humid; temperate winter. Sandy loams.
  Northern atlantic Morocco (Gharifa).
- *Frankenio pulverulentae-Suaedetum brevifoliae* Bendaanoun 1991
  Thermomediterranean. Arid, Semi-arid, Sub-humid; temperate, warm winter. Clayey loams, loams.
  Middle atlantic Morocco (Zima, Mellah) Northern atlantic Morocco (Bou Regreg, Loukkos, Gharifa) Mediterranean coast (Moulouya).
- *Limonio (delicatuli) leptostachydis-Suaedetum brevifoliae* Bendaanoun 1991
  Thermomediterranean. Semi-arid; temperate, warm winter. Loamy clays on sands.
  Mediterranean coast (Moulouya).
- *Limonio asparagoidis-Suaedetum brevifoliae* Bendaanoun 1991
  Thermomediterranean. Semi-arid; temperate, warm winter. Clayey loams, loams.
  Mediterranean coast (Nador).
- *Limonio ornati-Suaedetum brevifoliae* (Nègre 1960) Bendaanoun 1991
  Thermomediterranean. Arid; temperate winter. Loams.
  Middle atlantic Morocco (Zima).
- *Lycio intricati-Suaedetum brevifoliae* Bendaanoun 1991
  Thermomediterranean. Semi-arid; temperate winter. Clayey loams on sands.
  Middle atlantic Morocco (Mellah).
- *Mesembryanthemo nodiflori-Suaedetum brevifoliae* Bendaanoun 1991
  Infra-, Thermomediterranean. Arid, Semi-arid; temperate, warm winter. Loamy sands, loams, clayey loams.
  Middle atlantic Morocco (Souss, Oualidia) Mediterranean coast (Moulouya).
- *Sarcocornio alpini-Suaedetum brevifoliae* Bendaanoun 1991
  Infra-, Thermomediterranean. Arid, Semi-arid; temperate, warm winter. Clayey loams, loams.
  Middle atlantic Morocco (Souss, Tensift) Mediterranean coast (Moulouya).
- *Spergulario diandrae-Suaedetum brevifoliae* Nègre (1960) Bendaanoun 1991
  Thermomediterranean. Arid; temperate winter. Loams little limestone.
  Middle atlantic Morocco (Zima).
- *Sphenopo divaricati-Suaedetum brevifoliae* (Braun-Blanquet and Bolos 1958) Rivas-Martinez et al. 1983
  Thermomediterranean. Sub-humid; temperate, warm winter. Clayey loams.
  Northern atlantic Morocco (Bou Regreg, Gharifa, Loukkos).

- *Tamarico speciosi/boveanae-Suaedetum brevifoliae* Bendaanoun 1991
  Infra-, Thermomediterranean. Arid; temperate, warm winter. Sandy loams.
  Middle atlantic Morocco (Souss).

Suaedion fruticosae Nègre 1959

- *Lepturo incurvati-Suaedetum fruticosae* Nègre (1956) 1959
  Thermomediterranean. Arid; temperate winter. Loamy-clayey soils, salty (chlorures and sulfates), generally immersed for a short time of the year.
  Middle atlantic Morocco (Sedd El-Messjoun).
  Subassociations:

  – *betaetosum macrocarpae*
  – *lepturetosum incurvati*
  – *coronopetosum procumbensis*
  – *anabasetosum aphyllae*

- *Suaedo fruticosi-Limonietum ornati* Nègre (1956) 1959
  Thermomediterranean. Arid; temperate winter. Very sandy soils, salty (mainly chlorures) and saturated with water during a long period of the year.
  Middle atlantic Morocco (Haouz of Marrakech: Sedd El-Messjoun, Sidi Zouine, banks of Seguiet ech-Cherkaouia near Sidi Moulay-Barhdad. . .).

Coastal, Lagoon, Estuarian and Marine Ecosystems

Ammophiletea Br.-Bl. and Tüxen ex Westhoff, Dijk and Passchier 1946

(*Euphorbio paraliae-Ammophiletea australis* Gehu and Gehu-Franck 1988 corr. Gehu 1998)

Perennial communities; rhizomatous grasses and chamaephytes; sandy dunes of the seaboards.

Ammophiletalia *Br.-Bl. and Tüxen ex Westhoff, Dijk and Passchier 1946*

Ammophilion Br.-Bl. 1921

(*Ammophilion australis* Br.-Bl. 1933 corr. Rivas-Martinez, Costa and Izco in Rivas-Martinez et al. 1990)

- *Cypero kali-Euphorbietum paraliae* Sunding 1972

  Infra-, Thermomediterranean. Arid, Semi-arid; temperate, warm winter. Middle atlantic Morocco (both sides of Agadir, from the coastline of the Tiznit region (Sidi Moussa d'Aglou) to the north of Essaouira (Moulay Bouzerktour)).

  Most often found on the sandy veneer of cliffs with arenas; rarely in low dune systems.
- *Loto cretici-Elymetum farcti* Gehu and Sadki 1995

  Thermomediterranean. Semi-arid, Sub-humid; temperate, warm winter. Small embryonic dunes, most often clad or ascended on hilly sandy coastal terraces, or at the base of gravelly littoral ridges or the bottom of protected coves (Gehu and Biondi 1996).

M. S. Taleb, M. Fennane, *Vascular Plant Communities of Morocco*, Geobotany Studies, https://doi.org/10.1007/978-3-319-93704-5_7

Middle and Northern atlantic Morocco, between safi and Rabat, and maybe more towards the north.

Association floristically poor.

Subsassociations:

- *subassociation typicum*
- *crucianelletosum maritimae*
- *zygophylletosum fontanesii*

- *Otantho maritimae-Ammophiletum australis* Gehu and Tüxen 1975
  Thermomediterranean. Semi-arid; temperate winter. Sand, gravel.
  Middle atlantic Morocco (south Casablanca, north Safi, El Oualidia).
- *Polycarpaeo niveae-Crucianelletum maritimae* Gehu and Biondi 1996
  Thermomediterranean. Semi-arid; warm winter. Alveolar sandstones more or less sanded.
  Short portion of the coastline in the north of Essaouira (upper part of small cliffs).
- *Polycarpaeo niveae-Ammophiletum australis* Gehu and Biondi 1996
  Thermomediterranean. Semi-arid; warm winter.
  Middle atlantic Morocco (from south Essaouira until south of Safi).
- *Polycarpaeo niveae-Elymetum farcti* Gehu and Biondi 1996
  Infra-, Thermomediterranean. Arid, Semi-arid; temperate, warm winter. Sandy.
  Middle atlantic Morocco (between Safi and Agadir).
  The association occupies two particular microtopographic situations: small embryonic moving dunes at the top of the beach, or small dunes and light veneers of sand accumulated on the edges of small cliffs sprayed with "embruns".
- *Polycarpaeo niveae-Sporoboletum arenarii* Gehu and Biondi 1996
  Infra-, Thermomediterranean. Arid, Semi-arid; temperate, warm winter. Substrate formed of a thin layer of sand often encrusted on the surface, or of micro-dunes not thick, deposited on the sandy sandstone of the coastal terraces. Little sandy but very halophilous soil.
  Middle atlantic Morocco (from the Tiznit region, Sidi Moussa d'Aglou plage, until Moulay Bouzerktour in the north of Essaouira).
  Original association, certainly endemic.
  Subassociations:

  - *cakiletosum susicae*
  - *euphorbietosum paraliae*
  - *frankenietosum corymbosae*

- *Zygophyllo waterlotii-Euphorbietum paraliae* Gehu and Biondi 1996
  Saharan bioclimate; warm, temperate winter. Sandy soils.
  Saharan Morocco (south of Draa).
  Association observed and studied at Tan-Tan beach. Ecologically, it colonizes low dunes more or less moving, over a sandy beach.

Subassociations:

- *subassociation typicum*
- *traganetosum moquini*
- *mesembryanthemetosum cristallini*

Cakiletea maritimae Tüxen and Preising ex Braun-Blanquet and Tüxen 1952

Pioneer halo-nitrophilous annual communities of strandlines, beaches and coastal sand dunes.

Thero-Atriplicetalia *Pignatti 1953*

(*Euphorbietalia peplidis* Tuxen ex Rivas Goday et Rivas-Martinez 1958)

Euphorbion peplidis Tüxen ex Oberd. 1952

(*Cakilion maritimae* Pignatti 1953)

- *Cakiletum susicae* Gehu and Biondi 1996
  Infra-, Thermomediterranean. Arid, Semi-arid; temperate, warm winter. Sandy, nitrophilous.
  Remarkable association, endemic to the coastline from Essaouira to Tiznit, centered on Agadir.
  It develops at the tops of beaches, in estuarine borders, sometimes on the edges of the small sandy cliffs, or even present also in the hollows of more internal dunes.
- *Salsolo kali-Cakiletum maritimae* Costa and Mansanet 1981
  Thermomediterranean. Semi-arid; temperate, warm winter. Littoral sands.
  Middle atlantic Morocco (north of Safi).

Crithmo-Staticetea Br.-Bl. in Braun-Blanquet, Roussine and Nègre 1952

(*Crithmo-Limonietea* Br.-Bl. in Braun-Blanquet, Roussine and Nègre 1952)
Coastal rupicolous dwarf-herb communities; cliffs and lithosols splashed by marine salt spray.

Crithmo-Staticetalia *Molinier 1934*

Crithmo-Daucion halophili Rivas-Martinez, Lousa, Diaz, Fernandez Gonzalez and Costa 1990

- *Astidamio latifoliae-Chenoleetum tomentosae* Gehu and Biondi 1998

  Inframediterranean. Arid; warm winter. Mainly on the edges of plateaux and small cliffs exposed to sea spray.

  Substrates generally very enriched in various shell debris more or less old.

  Middle atlantic Morocco (Around Sidi Ifni, and further towards north until Tiznite).

  Association formed mainly by the canarian species *Astidamia latifolia* and *Chenolea tomentosa*.

  Subassociations:

  – *frankenietosum laevis*
  – *salsoletosum longifoliae*

- *Astidamio latifoliae-Suaedetum verae* Gehu and Biondi 1998

  Thermomediterranean. Semi-arid; warm winter. At the bottom of the coastal cliffs.

  Middle atlantic Morocco (Safi).

- *Frankenio laevis-Zygophylletum fontanesi* Gehu and Biondi 1998

  Infra-, Thermomediterranean. Arid, Semi-arid; temperate, warm winter.

  Uncommon, observed from Sidi Ifni in the south to Al Jorf and Oualidia in the North with a large hiatus on both sides of Agadir.

  It lives mainly at the foot of the cliffs, on the deposits of pebbles, in the crevices of rocks and alveolate slabs.

  Subassociations:

  – *subassociation typicum*
  – *atriplicetosum infniensis*

- *Leontodonto tingitani-Reichardietum picroides* Deil 1994

  Thermomediterranean. Sub-humid; warm winter. Coastal cliffs.

  Rif (region of Tanger).

- *Limonietum emarginati* Asensi 1984

  Thermomediterranean. Sem-arid; warm winter.

  Rif (between Sebta and Tetouan).

- *Limonio mucronati-Astidamietum latifoliae* Gehu and Biondi 1998

  Infra-, Thermomediterranean. Arid, Semi-arid; temperate, warm winter.

  Western Anti-Atlas and Middle atlantic Morocco (observed sporadically around Sidi Ifni/Mirhleft and Safi).

  This association lives at the base of the cliffs or on small vires and rocky escarpments in areas heavily sprayed with "embruns".

  Subassociations:

– *subassociation tpicum*
– *frankenietosum laevis*

Helianthemetea guttati (Br.-Bl.) Rivas-Goday and Rivas-Martinez 1963

(*Tuberarietea guttatae* (Br.-Bl.) Rivas-Goay and Rivas-Martinez 1963)
Ephemeral lowgrown communities (short herbs and grasses) on acidic substrates.

Vulpietalia *Pignatti 1953*

Ononidion tournefortii Gehu and Biondi 1996

- *Cypero kali-Ononidetum tournefortii* Wildpret et al. 1983
  Infra-, Thermomediterranean. Arid, Semi-arid; temperate, warm winter. Littoral dune areas semi-stabilized.
  Middle atlantic Morocco.
  The majority of the species of the association are annuals.
  Subassociations:

– *subassociation typicum*
– *salsoletosum kali*
– *rumicetosum picti*
– *polycarpaeetosum niveae*

- *Euphorbio paraliae-Ononidetum tournefortii* Gehu and Biondi 1996
  Inframediterranean. Arid, Saharan; warm winter.
  Semi-stabilized dune zones of Sidi Ifni and Tan-Tan beaches.

Malcolmietalia *Rivas Goday 1958*

- *Community of Aira uniaristata* Molina, Tahiri, Agostinelli et al. 2009
  Thermomediterranean. Semi-arid, Sub-humid; temperate, warm winter.
  Northern atlantic Morocco (Hinterland of the littoral between Kenitra and Mohammadia).

Polycarpaeo niveae-Traganetea moquini Rivas-Martinez and Wildpret in Rivas-Martinez et al. 2002

Saharan halophilous coastal and inland vegetation; tall and dwarf shrub permanent communities on sandy dunes and regosols.

Zygophyllo fontanesii-Polycarpaetalia niveae *Santos ex Géhu et al. 1996*

(*Ononidetalia ramosissimae* Galan de Mera, Sánchez Garcia and Vicente Orellana 1997)

Traganion moquinii Sunding 1972

(*Ononido ramosissimae-Polycarpion niveae* Biondi, Allegrezza, Taffetani and Wildpret 1994)

- *Cypero kali-Polycarpaeetum niveae* Gehu and Biondi 1996
  Infra-, Thermomediterranean. Arid, Semi-arid; warm, temperate winter. Sandy sandstone. Edges of small cliffs; sandy sandstone plateaux.
  Middle atlantic Morocco (From Tiznit to Safi: from Sidi Moussa d'Aglou in the South to Souria Kedima in the North).
  Subassociations:
  - *subassociation typicum*
  - *frankenietosum corymbosae*
- *Frankenio corymbosae-Ononidetum ramosissimae* Gehu and Biondi 1996
  Infra-, Thermomediterranean. Arid, Semi-arid; warm, temperate winter. Substrate somewhat eutrophic. Sandy soil.
  Middle atlantic Morocco (Sidi R'bat, Moulay Bouzerktour, Tifnite, Tamri).
  Developed on sands rather thick and little consolidated or in large dune systems. Its appearance coincides with areas more or less disturbed by human activities in the vicinity of fishing villages or crops.
- *Frankenio ericifoliae-Traganetum moquinii* Bendaanoun 1991
  Inframediterranean. Arid; warm winter. Sandy dunes.
  Middle atlantic Morocco (Massa).
- *Polycarpaeo niveae-Bubonietum imbricati* Gehu and Biondi 1996
  Infra-, Thermomediterranean. Arid, Semi-arid; warm, temperate winter.
  Middle atlantic Morocco (Sidi Moussa d'Aglou, Sidi Rbat Massa, Littoral Nord du Cap Rhir, Sidi Toual).

The association grows on large surfaces at the top of the slightly sloping and more or less sandy plateaux of the coastal sandy cliffs. Its stations are generally swept by sea winds.

Subassociations:

- *cyperetosum kali*
- *launaeetosum arborescentis*

- *Polycarpaeo niveae-Helichrysetum rhirense* Gehu and Biondi 1996
  Infra-, Thermomediterranean. Arid, Semi-arid; warm, temperate winter.
  The association occupies small areas, observed on both sides of the city of Agadir, from Sidi R'Bat to Tamri.
  It grows on the top of sandy cliffs, perhaps a little further back from sea, in areas still swept by winds but semi-protected.
- *Polycarpaeo niveae-Lotetum pseudocretici* Gehu and Biondi 1996
  Infra-, Thermomediterranean. Arid, Semi-arid; warm, temperate winter.
  Probably endemic to the maritim dunes of the Middle atlantic Morocco (Sidi R'bat, estuaire Massa, South Essaouira).
  Association developed in the zone behind dunes, sheltered from the winds, under the protection of Traganum moquini, in large dune systems.
- *Traganetum moquinii* Sunding 1972
  Inframediterranean. Arid, Saharan; warm winter. Sandy dunes.
  Common association for Canary islands and the littoral of Middle atlantic Morocco from Agadir to Tan-Tan.

Other groupements

- *Association with Andryala mogadorensis* Br.-Bl. and Maire 1924
  Thermomediterranean. Semi-arid; warm winter. Sandy soils.
  Middle atlantic Morocco (Essaouira island).
- *Association with Chenolaea canariensis* Br.-Bl. and Maire 1924
  Thermomediterranean. Semi-arid; warm winter. Sandy soils.
  Middle atlantic Morocco (Essaouira island).

Juncetea maritimae Br.-Bl. in Braun-Blanquet, Roussine and Nègre 1952

Perennial graminoid wet communities, dominated by rushes (*Juncus spp.*) and rhizomateous grasses. Coastal and inland temporary wet or inundated salt marshes.

Juncetalia maritimae *Br.-Bl. ex Horvatic 1934*

Juncion maritimae Br.-Bl. ex Horvatic 1934

- *Astero-Puccinellietum fasciculatae* (Beeftink 1965) Gehu 1976
  Thermomediterranean. Sub-humid; warm, temperate winter.
  Northern atlantic Morocco (Bou regreg) (Bendaanoun 1991:376).
- *Cynodo hirsutissimi-Juncetum maritimi* Bendaanoun 1991
  Thermomediterranean. Semi-arid, Sub-humid; temperate winter. Clayey loams on sands, sandy loams on sands.
  Middle atlantic Morocco (Sidi Moussa) Northern atlantic Morocco (Moulay Bou Selham).
- *Frankenio (corymbosae) laxae-Juncetum maritimi* Bendaanoun 1991
  Infra-, Thermomediterranean. Arid; warm winter. Sandy loams.
  Middle atlantic Morocco (Souss, Massa).
- *Inulo crithmoidis-Juncetum acuti* Bendaanoun 1991
  Thermomediterranean. Semi-arid; temperate, warm winter. Sandy loams, clayey loams on sands, sandy loams on sands.
  Mediterranean coast (Nekkor, Ghiss).
- *Limonio cymuliferi-Juncetum acuti* Bendaanoun 1991
  Thermomediterranean. Semi-arid; warm, temperate winter. Sands.
  Mediterranean coast (Nador).
- *Limonio ferulacei-Juncetum maritimi* Bendaanoun 1991
  Thermomediterranean. Semi-arid, Sub-humid; temperate, warm winter. Sandy loams.
  Middle atlantic Morocco (Oualidia) Northern atlantic Morocco (Tahadart, Gharifa) Rif (Martil, Smir).
- *Samolo valerandi-Juncetum maritimi* Fukarek 1961
  Thermomediterranean. Semi-arid; temperate winter. Clayey loams on sands, sandy loams on sands.
  Middle atlantic Morocco (Sidi Moussa, Oualidia).
- *Spergulario (marginatae) angustatae-Juncetum maritimi* Bendaanoun 1991
  Inframediterranean. Arid; warm winter. Clayey loams on sands.
  Middle atlantic Morocco (Massa).
- *Triglochino laxiflorae-Juncetum maritimi* Braun-Blanquet 1931
  Thermomediterranean. Semi-arid; warm winter. Sands.
  Rif (Smir).
- *Triglochino maritimi-Juncetum maritimi* Braun-Blanquet 1931
  Thermomediterranean. Sub-humid; temperate winter. Muddy sands, sandy mudds.
  Northern atlantic Morocco (Moulay Bou Selham).

Thero-Salicornietea Tüxen in Tüxen and Oberdorfer 1958

Littoral and inland halophytic communities of annual succulent plants. Mud flats, edges of the irregularly flooded salines, temporary inundated salt marshes and salt-pans.

Class very little known in Morocco; at least, the following associations may exist in our Mediterranean and Atlantic coastal regions (Bendaanoun 1991: 580), but we do not have more informations.

- *Arthrocnemo glauci-Salicornietum ramosissimae* Brullo and Firnari 1976
  Cf. Bendaanoun (1991:347–354 and 580).
- *Salicornietum ramosissimae* Gehu J.-M., Gehu J. and Caron 1978
  Cf. Bendaanoun (1991:347–354 and 580).
- *Salicornietum strictae* Christiansen 1955
  Cf. Bendaanoun (1991:347–354 and 580).

Zosteretea Pignatti 1953

Cormophyte benthic communities dominated by *Zostera*; littoral up to 5 m deep; muddy and sandy submerged substrates of lagoons, estuaries and calm coast shelves.

Zosteretalia *Béguinot ex Pignatti 1953*

Zosterion marinae Br.-Bl. and Tüxen ex Pignatti 1953

- *Zosteretum marinae* (Borgesen 1905) Harmsen 1936
  Muddy sands.
  Middle atlantic Morocco (Oualidia, Sidi Moussa) Northern atlantic Morocco (Loukkos, Moulay Bou Selham).
- *Zosteretum noltii* Harmsen 1936
  Sandy mudds, soft mudds, consolidated mudds.
  Atlantic coast.

Ruppietea maritimae Tüxen ex Den Hartog and Segal 1964

Submerged rooted herbaceous vegetation growing on temporary or permanent littoral or continental salts-water pools and lagoons.

Ruppietalia maritimae *Tüxen ex Den Hartog and Segal 1964*

Ruppion maritimae Br.-Bl. ex Westhoff in Bennema, Sissingh and Westhoff 1943

- *Ruppietum maritimae* (Warm. 1906) Hocq. 1927
  Mudds.
  Middle atlantic Morocco (Oualidia, Sidi Moussa, Mellah) Northern atlantic Morocco (Bou Regreg, Moulay Bou Selham, Tahadart) Mediterranean coast (Moulouya).

Spartinetea maritimae Tüxen in Beeftink and Gehu 1973

Pioneer vegetation of perennial cord grasses on tidal flats of temperate seas.

Spartinetalia maritimae *Conard 1935 ex Beeftink & Gehu 1973*

Spartinion maritimae Conard 1935 ex Beeftink & Gehu 1973

- *Bostrychio scorpioides-Spartinetum maritimae* Bendaanoun 1991
  Sandy mudds, mudds on sands.
  Northern atlantic Morocco (Bou Regreg).
- *Bostrychio scorpioides-Spartinetum townsendii* Bendaanoun 1991
  Sandy mudds.
  Northern atlantic Morocco (Bou Regreg).
- *Fuco axillaris-Spartinetum maritimae* Bendaanoun 1991
  Soft mudds.
  Middle atlantic Morocco (Oualidia, Sidi Moussa).
- *Fuco vesiculosi-Spartinetum maritimae* Bendaanoun 1991
  Consolidated sandy mudds, consolidated mudds.
  Northern atlantic Morocco (Loukkos).
- *Spartinetum maritimae* (Emb. and Reg. 1926) Corillion 1953
  Mudds.
  Middle atlantic Morocco (Oualidia, Sidi Moussa, Mellah) Northern atlantic Morocco (Bou Regreg, Sebou, Loukkos, Tahadart).
- *Suaedo (maritimae) perennantis-Spartinetum townsendii* Bendaanoun 1991
  Consolidated sandy mudds, consolidated mudds.
  Northern atlantic Morocco (Bou Regreg).

Spartinion junceae Bendaanoun 1991

- *Paspalo vaginati-Spartinetum densiflorae* Bendaanoun 1991
  Northern atlantic Morocco (Moulay Bou Selham).
- *Spartinetum densiflorae* Rivas-Martinez 1980
  Clayey loams.
  Northern atlantic Morocco (Moulay Bou Selham).
- *Spartinetum junceae patentis* Beeftink 1968 em. Bendaanoun 1991
  Clayey loams on sands.
  Rif (Smir).
- *Spartinetum junciformis* Bendaanoun 1991
  Mudds on sands, sandy mudds.
  Northern atlantic Morocco (Moulay Bou Selham).
- *Triglochino maritimi-Spartinetum junciformis* Bendaanoun 1991
  Sandy mudds.
  Northern atlantic Morocco (Moulay Bou Selham).

Saharan, Steppic or Pre-Steppic Vivace Communities (Shrublets and/or Graminaceous Herbs) and Annual or Perennial Vegetation (Incl. Adventices, Messicoles, Ruderals)

Pergulario tomentosae-Pulicarietea crispae Quézel 1965

Pergulario tomentosae-Pulicarietalia crispae *Quézel 1965*

Antirrhino ramosissimi-Zillion macropterae Quézel 1965

- *Plantago ciliatae-Ormenisetum lonadioidis* Quézel 1965
  Saharan bioclimate; temperate winter. 520–600 m. Limestone, clay, loam. Stony soils.
  Saharan Morocco (Hamada of Draa).
- *Rhantherio adpressi-Fagonietum zilloides* Quézel, Barbéro, Benabid and Rivas-Martinez 1994
  Saharan bioclimate; cool winter. Sandy soils.
  Saharan Morocco (between Errachidia and Tineghir; lower valley of Draâ). Small sandy depressions, margins of bed rocky wadis.
- *Zizipho loti-Acacietum raddianae* Quézel 1965
  Saharan, Arid; temperate to cool winter. 300–700 m. Alluvions of loamy sand; deep soils.
  Saharan Morocco (Oued Mird, southern hamadas, jbel Bani, south flank of the Saghro) (Fig. 1).
- *Zizipho loti-Retametum sphaerocarpae* Coquillard 1982
  Saharan, Arid; temperate to cool winter. Sandy, rocky soils.
  Saharan Morocco (Tafilatet).

Panico turgidi-Acacion raddianae Quézel (1954) 1965 (Fig. 2)

- *Balanito aegyptiacae-Acacietum ehrenbergianae* Quézel, Barbéro, Benabid and Rivas-Martinez 1995

M. S. Taleb, M. Fennane, *Vascular Plant Communities of Morocco*, Geobotany Studies, https://doi.org/10.1007/978-3-319-93704-5_8

Fig. 1 *Zizipho loti-Acacietum raddianae*, South of Morocco (ph. Taleb)

Saharan; temperate winter. 250–450 m. Alluvions, silty-sandy; little deep soils.

Saharan Morocco (jbel Ouarkziz and more toward the south).

- *Cassio-Panicetum turgidae* Quézel 1965

  Saharan; temperate winter. 300–1800 m. Coarse alluvium; soils rocky or gravelly.

  Saharan Morocco.

  Heterogeneous association, showing a great variation concerning its floristic composition.

- *Foleyolo billotii-Acacietum raddianae* Quézel, Barbéro, Benabid and Rivas-Martinez 1995

  Saharan; temperate winter. 300–700 m. Silty-sandy; stony soils.

  Beds of the Saharan rivers (Fig. 3).

- *Psoraleo plicatae-Hyoscyametum mutici* Quézel 1965

  Saharan; temperate winter. 200–750 m. Silty or silty-sandy.

  Saharan Morocco (river beds, sandy depressions).

- *Rottboellio hirsutae-Maeruetum crassifoliae* Quézel, Barbéro, Benabid and Rivas-Martinez 1995

  Saharan; temperate winter. 400–550 m. Sandy, sandy rocky.

  Saharan Morocco (region between Foum el Hisn and Akka).

- *Stipagrostido pungentis-Acacietum raddianae* Quézel, Barbéro, Benabid and Rivas-Martinez 1995

  Saharan; temperate winter. 350–550 m. Sandy.

Fig. 2 Saharan steppe with *Acacia raddiana*, Oued Mird, South of Morocco (ph. Taleb)

Fig. 3 *Foleyolo billotii-Acacietum raddianae*, Oued Mird, South of Morocco (ph. Taleb)

Fig. 4 *Stipagrostido pungentis-Acacietum raddianae*, Oued Mird, South of Morocco (Ph. Taleb)

Saharan Morocco (south of jbel Ouarkziz, region of Zagora and M'hamid) (Fig. 4).

Asterisco graveolensis-Forskahletea tenacissimae Quézel 1965

Gymnocarpo decandri-Atractyletalia babelii *Quézel 1965*

Atractylion babelii Lemée 1952

- *Fagonietum harpago-longispinae* Quézel, Barbéro, Benabid and Rivas-Martinez 1995

  Saharan; temperate winter. 500–800 m. Quartzite; rocky substrate.

  Saharan Morocco / Anti-Atlas (Draa basin: meridional portion of the Anti-Atlas between Aouinet Torkoz and Akka).
- *Fredolietum aretioidis* Quézel 1965

  Mesomomediterranean. Saharan, lower Arid; cool to cold winter. 350–1650 m. Limestone, marly-calcareous; stony soils.

Fig. 5 *Fredolietum aretioidis*, Region of Errachidia (ph. Taleb)

Eastern High-Atlas (Ziz valley between Rich and Amouguer), Op (eastern high plateaux) (Figs. 5 and 6).

- *Morettio-Fagonietum longispinae* Lemée 1953
  Saharan, Arid; cool to cold winter. Rocky.
  South region of Saharan Atlas.
  Subassociations:

  – *subassociation typicum*
  – *anabasidetosum aretioides*
  – *aristidetosum floccosae*
  – *artemisietosum herbae-albae*

- *Withanio adpressae-Linarietum sagittatae* Quézel 1965
  Thermo- and Mesomediterranean. 500–1500 m. Saharan, Arid; temperate to cool winter. Rocky.
  Saharan Morocco/Anti-Atlas (jbel Bani, Tafilalet, Foum El-Hassane, Hamadas,. . .), High-Atlas (south flank).

Fig. 6 *Fredolia aretioides* (ph. Taleb)

Sedetalia altissimae *Lemée 1953*

- *Ferulo cossonianae-Leucanthemetum gaetulae* Lemée 1953
  Thermo- and Mesomediterranean. 850–1600 m. Saharan, Arid; temperate to cool winter. Limestone, clay; rocky soils.
  Anti-Atlas, Saharan Atlas (jbel Zenaga).
  Subassociations:

  – *aristidetosum adscensionis*
  – *rosmarinetosum officinalis*

Senecietalia flavi *Quézel 1965*

Senecion flavi Quézel 1965

- *Centaureo-Senecietum flavi* Lemée 1953
  Saharan, Arid; cool winter. 890–1050 m. Rocky.
  South region of Saharan Atlas (jbel Zenaga...).
- *Enarthrophyto-Tourneuxietum variifolii* Quézel 1965
  Saharan, Arid; temperate to cool winter. Rocky substrtate.
  Southern Morocco.

- *Maeruo-Ephedretum rollandii* Quézel 1965
  Saharan; temperate winter. 600–700 m.
  Saharan Morocco (mountain massif of Zemmour).
- *Danthonio forskahli-Plantagonetum ciliatae* Quézel 1965
  Saharan; temperate winter.
  Saharan Morocco.

Cutandietea?

Unknown class; informations extracted from Lemée (1953:143).

- *Trigonello-Althaeetum ludwigii* Lemée 1953
  Saharan, Arid; temperate to cool winter. Loamy cuvettes.
  Saharan Morocco/Saharan Atlas (Algerian frontier in the region of Figuig).
- *Lotononidetum dichotomae* Lemée 1953
  Saharan, Arid; temperate to cool winter. Sandy small ravines.
  Saharan Morocco/Saharan Atlas (Algerian frontier in the region of Figuig).
- *Asphodeletum pendulini* Lemée 1953
  Saharan, Arid; temperate to cool winter. Windy sandy short slopes.
  Saharan Morocco/Saharan Atlas (Algerian frontier in the region of Figuig).

Other Groupements

- *Peganeto-Bassietum muricae* Lemée 1953
  Saharan, Arid; temperate to cool winter. Soils rich in nitrogen.
  Saharan Morocco/Saharan Atlas (Algerian frontier in the region of Figuig).
  Subassociations:
  - *trigonelletosum anguinae*
  - *asphodeletosum pendulini*
- *Association with Warionia saharae* Lemée 1953
  Saharan, Arid; temperate to cool winter. Calcareous and sandstone rocky substrate.
  Saharan Morocco/Saharan Atlas (Algerian frontier in the region of Figuig: Antar and Zenaga mountains).
  Chasmophytic association.
- *Association with Notholaena vellea and Polygala rupestris var. saxatilis* Lemée 1953
  Saharan, Arid; temperate to cool winter. Above 1400 m altitude. Limestone, rocky substrate. North exposure.
  Saharan Morocco/Saharan Atlas (Algerian frontier in the region of Figuig: Antar mountain).
  Chasmophytic association.

Lygeo sparti-Stipetea tenacissimae Rivas-Martinez 1978

(*Thero-Brachypodietea ramosi* Br.-Bl. ex A. Bolòs & O. Bolòs in Bolòs and Vayerda 1950)

Tall or short herbaceous vegetation, annual or perennial, on sandy-loamy soils over calcareous bedrocks.

Very little known in Morocco and South and East Mediterranean.

Cymbopogono-Brachypodietalia ramosi *Horvatić 1963*

(*Thero-Brachypodietalia* Br.-Bl. 1931)

(*Hyparrhenietalia hirtae* Rivas-Martinez 1978)

- *Asphodelo tenuifoliae-Spergularietum albae* Nègre 1956
  Thermomediterranean. Arid; temperate winter. Loamy soils, more or less clayey.
  Middle atlantic Morocco (Sedd El-Messjoun: WNW of the Azib bel Harek).
- *Centaureo pungensi-Cynodonetum dactyli* Nègre 1959
  Thermomediterranean. Arid; temperate winter. Loamy, loamy-clayey.
  Middle atlantic Morocco (western Haouz: oued Amejnez north-western of Imi-n-Tanoute).
- *Chenopodio murali-Sonchetum maritimae* Peltier 1982
  Thermomediterranean. Arid; temperate winter. 50–300 m. Soil with more or less silty texture with sand accumulation on the surface.
  Middle atlantic Morocco (Souss plain).
- *Silenetum abietum-cuatrecasasii* Quézel, Barbéro, Benabid, Loisel and Rivas-Martinez 1988
  Mesomediterranean. 950–1540 m. Sub-humid, Humid; cool winter. Dolomite.
  Rif mountains.

Lygeo-Stipetalia tenacissimae *Br.-Bl. and O. de Bolos 1958*

(*Stipetalia tenacissimae* Kaabeche (1990) 1996)

Arid and Semi-arid grassland communities dominated by *Stipa tenacissima*, *Lygeum spartum, Artemisia herba-alba, Noea mucronata*... growing on calcareous and clayey soils without any temporary hydric process or salinity. (Fig. 7).

Up to now, two alliances are konwn in North Africa, ***Stipo tenacissimae-Launeion acanthocladae*** Kaabeche (1990) 1996 and ***Noaeo mucronatae-Artemision herba-albae*** Aïoud-Lounis 1984; all the associations need to be studied. We think that the followings ones, described by Peltier (1982) on the high plateaux of the western Anti-Atlas, belong to this order; Peltier has approved this point of view by email in September 2017.

Fig. 7 Steppe of *Stipa tenacissima*, eastern High-Plateaux (ph. Taleb)

- *Artemisio incultae-Convolvuletum trabutiani* Peltier 1982

  Mesomediterranean. 1250–1600 m. Semi-arid, Arid; cool, cold winter. Limestone, dolomite. Shallow loamy soils, generally with a fine to very fine texture.

  Anti-Atlas (Adrar Fouilim, adrar Minount, jbel Tanchkirra, jbel Taghzout) (Fig. 8).
- *Artemisio incultae-Erodietum jahandieziani* Peltier 1982

  Mesomediterranean, Mountain mediterranean. Semi-arid; cold winter. 1650–1850 m. Dolomite, quartzite, conglomerate.

  Median and western parts of high plateaux of the Anti-Atlas (Iberkâkène, Tanloût, hills of Tingarf).

  Subassociations:

  – *subssociation with Lavandula dentata, Thymus satureioides and Ononis natrix*

  – *subassociation with Thymus pallidus and Chamaerops humilis*
- *Artemisio incultae-Erodietum meynieri* Peltier 1982

  Mesomediterranean. 1100–1400 m. Arid, Semi-arid; cool, cold winter. Dolomite, limestone. Silty-clayey or clayey soils.

  Anti-Atlas (hills of Timkit; region of Imi-n-Oumghaz and Azaghar n-Iris; first slopes of adrar Tirrout and jbels Taghzout and Tasguelt).
- *Artemisio incultae-Stipetum parviflorae* Peltier 1982

  Mesomediterranean. 1100–1400 m. Arid, Semi-arid; cool, cold winter. 700–1750 m. Quartzites, rhyolites, siliceous conglomerate. Lithosols ou regosols.

Fig. 8 *Artemisio incultae-Convolvuletum trabutiani*, Anti-Atlas (ph. Taleb)

Anti-Atlas (regions of Kerdous; Idraren, Ou-Oumerreksou, Ighir-ou-Drar, Ighchène, jbel Toungoute, jbel Tagwilalt).

- *Artemisio incultae-Thymetum satureioidis* Peltier 1982

  Mesomediterranean, Mountain mediterranean. Semi-arid; cold winter. 1550–2050 m. Quartzites, rhyolites, siliceous conglomerate. Deep soil, not or little carbonate.

  Steppic areas in the south east of the Anti-Atlas (Azaghar-n-Yelim, Adrar n-Fou-Zagharn) and the neighbouring slopes of adrar Amjoud, jbel Agouti, jbel Iguiguil, jbel Aklim, adrar Minount (Fig. 9).

Tillaeetea muscosae Nègre (1956) 1959

Phytosociological class identified by Nègre (1959) on calcareous skeletal soils in the Northern part of the Middle atlantic Morocco.

Fig. 9 *Artemisio incultae-Thymetum satureioidis*, Anti-Atlas (ph. Taleb)

Spergularietalia fimbriatae *Nègre 1959*

- *Calendulo arvensis-Diplotaxetum assurgensis* Nègre 1959
  Thermomediterranean. Arid, Semi-arid; temperate winter.
  Small places included in the area of the *Notocero bicorni-Plantaginetum ovatae* (cf. infra).
- *Eruco sativae-Astragaletum maroccanae* Nègre 1959
  Thermomediterranean. Arid; temperate winter.
  Middle atlantic Morocco (north of Marrakech).
- *Lycio intricati-Plantaginetum ovatae* Nègre 1959
  Thermomediterranean. Arid, Semi-arid; temperate winter. Limestone, sandy loams. Stony.
  Middle atlantic Morocco (Sidi Zouine; western Haouz between the road Chichaoua/Marrakech and Oued Tensift).
- *Plantago coronopi-Vagarietum legionarii* Nègre 1959
  Thermomediterranean. Arid, Semi-arid; temperate winter. Soil more or less sandy and slightly calcareous on the surface, rich in clays and more limestone in depth.
  Middle atlantic Morocco (western Haouz, west and south-west of Chichaoua).
  Subassociations:

– *ranunculusetosum bullati*
– *hannonietosum hesperidi*

- *rumexetosum bucephalophori*
- *arisarumetosum simorrhinumi*

- *Rumexo bucephalophori-Plantaginetum coronopiae* Nègre 1959
  Sandy soils, siliceous, very little limestone.
  Thermomediterranean. Arid, Semi-arid; temperate winter. Middle atlantic Morocco (Hills bordering the north of the Beni Amir plain and the western Bahira until the West of the road Marrakech-Casablanca).
  Resulting directly from degradation of the Ziziphus and Chamaerops communities under the effect of grazing.
  Subassociations:

  - *androcymbietosum graminae*
  - *poeetosum bulbosae*

- *Stipo retortae-Notoceretum bicornae* Peltier 1982
  Infra- or? Thermomediterranean. Arid; warm, temperate winter. Quaternary alluvial sediments; more or less loamy.
  Middle atlantic Morocco (Upstream part of the Oued Souss valley).
- *Stipo retortae-Cladanthetum arabicae* Nègre 1959
  Thermomediterranean. Arid; temperate winter.
  Middle atlantic Morocco (Bni Moussa in the north of Marrakech)

Spergularietalia purpureae *Nègre 1959*

- *Aizoo hispanicae-Spergularietum purpurae* Nègre (1956) 1959
  Thermomediterranean. Arid; temperate winter. Lakeside limestone.
  Middle atlantic Morocco (Sedd El-Messjoun).
- *Androcymbio graminae-Irisetum sisyrinchiae* Nègre 1959
  Thermomediterranean. Arid; temperate winter. Alluvium, sandy and loamy; limestone.
  Middle atlantic Morocco (north and west of the Bni-Amir plain).
- *Androcymbio graminae-Notoceretum bicornae* Nègre 1959
  Thermomediterranean. Arid; temperate winter. Loams. Skeletal soils.
  Middle atlantic Morocco (loamy plain of Bni-Amir).
- *Artemisio herba-albae-Plantaginetum ovatae* Nègre 1959
  Thermomediterranean. Arid; temperate winter. Sandy loams; little thick soils.
  Middle atlantic Morocco (western Haouz).
- *Haloxylo scopariae-Plantaginetum ovatae* Nègre 1959
  Thermomediterranean. Arid; temperate winter. Siliceous; little thick soils.
  Middle atlantic Morocco (western Haouz: vast area around the jbel Ardouze; between Chichaoua and Imi-n-Tanoute).

- *Notocero bicorni-Astragaletum maroccani* Nègre 1959
  Thermomediterranean. Arid; temperate winter. Alluvium more or less thick, limestone and siliceous.
  Middle atlantic Morocco (Haouz: central Hadra).
- *Notocero bicorni-Plantaginetum ovatae* Nègre 1959
  Thermomediterranean. Arid; temperate winter. Phosphatic limestone, crusted.
  Middle atlantic Morocco (road Marrakech—El Jadida).
- *Spergulario purpurae-Hannonietum hesperidae* Nègre 1959
  Thermomediterranean. Arid; temperate winter. Sandy loams, clayey enough, on a limestone slab. Stony soil.
  Middle atlantic Morocco (south oued Oum Rbiâ: Bni-Moussa and Bni-Amir).
  Subassociations:

  – *hannonietosum hesperidae*
  – *sedumetosum caespitosae*
  – *notoceretosum bicorni*

- *Stipo retortae-Silenetum adustae* Nègre 1959
  Thermomediterranean. Arid; temperate winter. Stony loams.
  Middle atlantic Morocco (Bni Moussa near Oum-Rbiâ, Bni-Amir, towards the east until the hills that connect the Middle-Atlas to jbilete).
  Subassociations:

  – *irisetosum sisyrinchiae*

- *Stipo retortae-Resedetum myriospermae* Nègre 1959
  Thermomediterranean. Arid; temperate winter. Limestone slab. Soil rich in calcareous loams.
  Middle atlantic Morocco (Bahira; frequent in north and west of Sedd el-Messjoun).
- *Tillaeetum muscosae* Nègre 1959
  Thermomediterranean. Arid; temperate winter.
  Middle atlantic Morocco (north of Marrakech).
  Micro-association, with nanotherophytes.
- *Tunico illyricae-Spergularietum fimbriatae* Nègre (1956) 1959
  Thermomediterranean. Arid; temperate winter. Stony brown loams on limestone slabs.
  Middle atlantic Morocco (central Bahira in the north of Marrakech).

Notoceretea bicorni Nègre (1956) 1959

Soils generally well-permeable, not very thick, without limestone at least up to 50 cm deep.

Caracteristic species (Nègre 1977:26): *Aizoon canariense, Leysera leyseroides, Notoceras bicorne, Picris coronopifolia subsp. albida, Plantago amplexicaulis.*

Calenduletalia murbeckii ***Nègre 1959***

Stirred or cultivated schistous soils.

- *Malcolmio patulae-Calenduletum arvensis* Nègre 1959
  Thermomediterranean. Arid; temperate winter.
  Similar to *Iflogo-Calenduletum*, but more rich with ruderal species (*Diplotaxis spp.*), *Peganum harmala* abundant.
- *Wahlenbergieto nutabundae-Leyseretum leyseroidis* Nègre 1959
  Thermomediterranean. Arid, Semi-arid; temperate winter.
  Middle atlantic Morocco (western Haouz).
- *Aizoo hispanicae-Picridetum coronopifoliae* Nègre 1959
  Thermomediterranean. Arid; temperate winter.
  Limestone crusts; lightly salted soils.
  Middle atlantic Morocco (Haouz: small areas aroud Sidi Zouine).
- *Asterisco pygmae-Salsoletum sieberi* Nègre 1959
  Thermomediterranean. Arid; temperate winter. Permo-triasic sandstone.
  Middle atlantic Morocco (Bled El-Harousia north of Oued Tensift between Chichaoua and Chemmaîya).
  Subassociations:

  – *stipetosum retortae*
  – *frankenietosum corymbosae*

- *Filago heteranthae-Calenduletum algeriensis* Nègre 1959
  Thermomediterranean. Arid, Semi-arid; temperate winter. Mixture of calcareous and non-calcareous elements with granite dominance; deep cultivated soils.
  Middle atlantic Morocco.
- *Haloxylo scopariae-Aizoetum hispanicae* Nègre 1959
  Thermomediterranean. Arid; temperate winter. Terrace more or less conglomeratic in depth, not or little calcareous.
  Middle atlantic Morocco (South of oued Tensift, downstream of the confluence with oued Nfiss).
- *Notocero bicorni-Linarietum punctatae* Nègre 1959
  Thermomediterranean. Arid; temperate winter.
  Middle atlantic Morocco (small places in the borders of oued El-Abid).
- *Plantago ovatae-Calenduletum algeriense* Nègre 1959
  Thermomediterranean. Arid, Semi-arid; temperate winter. Mixture of calcareous and non-calcareous elements, but predominantly granitic deposited on a calcareous crust.
  Middle atlantic Morocco.

Haloxyletalia scopariae *Nègre (1956) 1959, 1977*

Non calcareous and non sandy soils: schists, loams...; not too deep.

Caracteristic species (Nègre 1977:27): *Haloxylon scoparium, Scorzonera undulata, Rumex vesicarious, Salvia aegyptiaca, Wahlenbergia lobelioides subsp. nutabunda, Aristida coerulescens.*

According to Nègre (1977), many alliances may exist within this order beside the ***Aizoetion canariensis*** Nègre 1959.

- *Calendulo (arvensis) hydruntinae-Matthioletum parviflorae* Nègre 1959
  Thermomediterranean. Arid; temperate winter. Essentially limestone, rich in loam, sand more or less granitic; skeletal soil.
  Middle atlantic Morocco (Jbilete).
- *Haloxylo scopariae-Erucetum longirostris* Nègre 1959
  Thermomediterranean. Arid; temperate winter. Schists.
  Middle atlantic Morocco (Hadra centrale).
- *Haloxylo scopariae-Leyseretum leyserioidis* Nègre 1959
  Thermomediterranean. Arid; temperate winter. Loamy, not calcareous. Stony.
  Middle atlantic Morocco (western Hadra, Lower parts of the northern flank of Jbiletes).
- *Haloxylo scopariae-Matricarietum maroccanae* Nègre 1959
  Thermomediterranean. Arid; temperate winter.
  Middle atlantic Morocco (Haouz: Sidi Zouine).
- *Haloxylo scopariae-Matthioletum parviflorae* Nègre 1959
  Thermomediterranean. Arid; temperate winter. Siliceous alluvium, sandy loam. Thalweg and shallow valley bottoms.
  Middle atlantic Morocco (north western Hadra).
- *Haloxylo scopariae-Spergularietum purpurae* Nègre (1956) 1959
  Thermomediterranean. Arid; temperate winter. Non-limestone deposited on lacustrine limestone.
  Middle atlantic Morocco (Sedd el-Messjoun).
- *Iflogo spicati-Aizoetum canariensis* Nègre 1959
  Thermomediterranean. Arid; temperate winter. Schistous loams, sandy clays.
  Middle atlantic Morocco (Summits and slopes of the Sidi Zouine hills, south slope of Jbilete between Oued N'fiss and the road Casa-Marrakech).
- *Iflogo spicati-Leyseretum leyserioidis* Nègre 1959
  Thermomediterranean. Arid; temperate winter. Loams predominant; deep soil, stony.
  Middle atlantic Morocco (central Haouz western of Marrakech).
- *Leysero leyseroidi-Mesembryanthemetum nodiflori* Nègre (1956) 1959
  Thermomediterranean. Arid; temperate winter. Loams; brown-red soil, eroded, stony.
  Middle atlantic Morocco (Bahira).

- *Leysero leyseroidi-Salsoletum vermiculatae* Nègre 1959
  Thermomediterranean. Arid; temperate winter. Clayey loams rich in stones.
  Middle atlantic Morocco (Jbilete).
- *Notocero bicorni-Leyseretum leyserioidis* Nègre 1959
  Thermomediterranean. Arid; temperate winter. Acid sandstone, loams. Stony.
  Middle atlantic Morocco (Hadra).
  A very homogeneous association, presents a single variant with Aizoon canariense, intermediate between Notocero-Leyseretum and Iflogo-Aizoetum.
- *Notocero bicorni-Mesembryanthemetum nodiflori* Nègre 1959
  Thermomediterranean. Arid; temperate winter. Schistous sandstone.
  Middle atlantic Morocco (Sidi Zouine).
- *Salvio aegyptiacae-Picridetum coronopifoliae* Nègre 1959
  Thermomediterranean. Arid; temperate winter. Clayey, sandy, loamy.
  Middle atlantic Morocco (eastern Jbiletes in the ouest of Sidi Zouine meridian).
  Subassociations:

– *wahlenbergietosum nutabundae*
– *andropogenotosum ischaemumi*
– *scorzoneretosum undulatae*
– *asteriscetosum pygmaei*
– *allietosum paniculati*

Ormenietalia mixtae *Nègre (1956) 1959, 1977*

Sandy soils.

- *Bromo rigidi-Sclerosciadetum nodiflorae* Peltier 1982
  Infra- or? Thermomediterranean. Arid; warm, temperate winter. Sand of marine origin; deep soil.
  Middle atlantic Morocco (Chtouka plain).
- *Evacido pygmae-Leyseretum leyserioidis* Nègre 1959
  Thermomediterranean. Arid; temperate winter. Limono-psammophile, calcarifuge; big non-calcareous elements.
  Middle atlantic Morocco (central Haouz south of Marrakech).
- *Iflogo spicati-Filagetum heteranthae* Nègre 1959
  Thermomediterranean. Arid; temperate winter. Roughly pure granite arena; thin soils (< 10 cm).
- *Limonio beaumieranae-Malcolmietum broussonnetii* Peltier 1982
  Infra- or? Thermomediterranean. Arid; warm, temperate winter. Limestone. Sandy.
  Middle atlantic Morocco (Plain of Chtouka).

- *Ormenio mixtae-Malcolmietum patulae* Nègre 1959
  Thermomediterranean. Arid; temperate winter. Granitic sands.
  Middle atlantic Morocco (granitic cuvette of Douar Ouaslem between main chain of Jbiletes and jbel Rass el-Feja).

Calenduletea algeriensis Nègre 1959

Herbaceous groupements; wasteland and crops; clay-limestone soils. Arid and Semi-arid bioclimates.

Class considered as synonym of *CHENOPODIETEA* by Mucina et al. 2016.

Calenduletalia algeriensis *Nègre 1959*

Soils "chatains" brown or "chatains" red.

Ceratocnemion rapistroidis Nègre 1977

- *Calendulo algeriensis-Allietum nigrii* Nègre 1959
  Thermomediterranean. Arid; temperate winter.
  Middle atlantic Morocco (east of Chemmaia).
- *Calendulo algeriensis-Psychinetum stylosae* Nègre (1956) 1959
  Thermomediterranean. Arid; temperate winter. Clayey.
  Middle atlantic Morocco (Bni-Amir, Bni-Moussa, eastern Haouz, Bahira).
  Subassociations:
  - *cladanthetosum arabicae*
  - *ceratocnemetosum rapistroidis*
  - *psychinetosum stylosae*
  - *volutarietosum lipii*
- *Scolymo maculati-Rhagadioletum stellati* Nègre 1959
  Thermomediterranean. Arid; temperate winter.
  Middle atlantic Morocco (eastern Haouz).

Diplotaxidetalia assurgentis *Nègre 1977*

Gray and brown soils, calcareous or phosphatic.

Volutarion crupinoidis Nègre 1977

- *Anacyclo valentinae-Diplotaxidetum tenuisiliquae* Nègre 1959
  Thermomediterranean. Arid; temperate winter. Loamy soils.
  Middle atlantic Morocco (western Haouz).
- *Atriplicito halimi-Calenduletum algeriensis* Nègre (1956) 1959
  Thermomediterranean. Arid; temperate winter. Loams.
  Middle atlantic Morocco (Bahira, central Haouz).
  Subassociations:

  – *calenduletosum algeriensis*
  – *anacyletosum valentini*

- *Atriplicito halimi-Trigonelletum anguinae* Nègre 1959
  Thermomediterranean. Arid; temperate winter.
  Middle atlantic Morocco (bled El-Harouzia northern oued Tensift).
  Quite similar to the *Atriplicito halimi-Calenduletum algeriensis.*
- *Eruco sativae-Calenduletum algeriensis* Nègre 1959
  Thermomediterranean. Above 450 m. Arid; temperate winter. Very silty, fairly clayey, formed on the limestones. Red brown soils.
  Middle atlantic Morocco (Hadra, western Haouz).
  Subassociations:

  – *calenduletosum*
  – *fumarietosum*
  – *tillaeetosum*

- *Fumario parviflorae-Diplotaxetum tenuisiliquae* Nègre (1956) 1959
  (*Lycio intricati-Asparagetum stipularis* Nègre 1956 subass. *fumarietosum*) (cf. Nègre 1959:227)
  Thermomediterranean. Arid; temperate winter. Loamy, loamy-sandy, not limestone.
  Middle atlantic Morocco (Bahira, south Jbilete in the eastern Haouz).
- *Iriso sisyrinchiae-Anabasetum aphyllae* Nègre (1956) 1959
  Thermomediterranean. Arid; temperate winter. Limestone. Brown, steppic soils.
  Middle atlantic Morocco (Northern border of the Bahira at the foot of the limestone hills of the Hadra).

Fig. 10 *Lycio intricati-Asparagetum stipularis* (ph. Taleb)

Subassociations:

– *diploschistesetosum scruposi*
– *calenduletosum algeriensis*
– *mesembryanthemetosum nodiflori*

- *Lycio intricati-Asparagetum stipularis* Nègre (1956) 1959
  Thermomediterranean. Arid; temperate winter. Siliceous, loams.
  Middle atlantic Morocco (Sedd el-Messjoun, western Haouz) (Fig. 10).
  Subassociations:

– *meliletosum sulcatae*
– *asphodelusetosum tenuifolii*
– *haloxyletosum scopariae*

- *Salsolo vermiculatae-Aizoetum hispanicae* Nègre 1959
  Thermomediterranean. Arid; temperate winter. Limestone.
  Middle atlantic Morocco (western Haouz).
  Subassociations:

– *calenduletosum*
– *linarietosum*
– *lycietosum*
– *mesembryanthemetosum*
– *plantaginetosum*

- *Salsolo vermiculatae-Frankenietum corymbosae* Nègre (1956) 1959
  Thermomediterranean. Arid; temperate winter. Loams. Phosphatic soils.
  Middle atlantic Morocco (Sedd el-Messjoun, western Haouz).
  Subassociations:

– *linarietosum bipartitae*
– *mesembryanthemetosum nodiflorae*
– *phalaridetosum minori*

Vicion monanthae Nègre 1977

- *Hypeco pendulae-Silenetum vulgaris* Nègre 1959
  Thermomediterranean. Semi-arid; temperate, cool winter. Siliceous. Phosphatic soils.
  Middle atlantic Morocco (around Imi-n-Tanout and the north High-Atlas border).
  Homogeneous association, with no significant variations.
- *Vicio biflorae-Silenetum vulgaris* Nègre 1959
  Thermomediterranean. Arid; temperate winter. Limestone. Phosphatic soils.
  Middle atlantic Morocco (cuvette of Sidi Mokhtar, cuvette of Tarselt).
  Subassociations:

– *medicagetosum litoralis*
– *caucalo leptophyllae-vaccarietosum pyramidae*
– *biaretosum (bovei) blumei*
– *matricarietosum (pubescentis) maroccanae*
– *galiumetosum tricorni*

- *Erodio guttati-Medicaginetum litoralis* Nègre 1959
  Thermomediterranean. Arid; temperate winter.
  Middle atlantic Morocco (small surfaces south eastern jbel Kharouba in the west of the road Chichaoua /Imi-n-Tanout).

Other groupements

- *Groupement with Eruca sativa var. longirostris* Br.-Bl. and Maire 1924
  Thermomediterranean. Arid; temperate, cool winter.
  North eastern Morocco (region of Oujda).

Cladanthetalia arabicae *Nègre (1956) 1977*

Soils little limestone, “chatains-bruns” or “brun-rouges”.

Caracteristic species (Nègre 1977:29): *Cladanthus arabicus, Diplotaxis virgata, Fumaria densiflora, Linaria micrantha, Silene muscipula.*

Diplotaxidion virgatae Nègre 1977

- *Aizoo hispanicae-Adonidetum dentatae* Nègre 1959
  Thermomediterranean. Arid; temperate winter. Limestone.
  Middle atlantic Morocco (Sidi Zouine).
- *Cladantho arabici-Calenduletum algeriense* Nègre 1959
  Thermomediterranean. Arid; temperate winter. Loamy-clayey.
  Middle atlantic Morocco (Haouz).
  Subassociations:

  – *rhagadioletosum stellati*
  – *pallenidetosum spinosae*

- *Cladantho arabici-Silenetum muscipulae* Nègre 1959
  Thermomediterranean. Arid, Semi-arid; temperate winter. Clays, loams. Phosphatic soils.
  Middle atlantic Morocco (cuvettes, glaciers with slight slopes bordering the phosphate massifs in western Chichaoua).
- *Fumario densiflorae-Diplotaxidetum virgatae* Nègre 1959
  Thermomediterranean. Arid; temperate winter. Siliceous. Phosphatic.
  Middle atlantic Morocco (Bahira).
- *Fumario densiflorae-Silenetum muscipulae* Nègre 1959
  Thermomediterranean. Arid; temperate winter. Phosphatic, sandy loam in the upper horizon, clayey-silty in depth.
  Middle atlantic Morocco (western Haouz, western Bahira).
- *Launaeo nudicauli-Volutarietum lipii* Peltier 1982
  Infra- or? Thermomediterranean. Arid; warm, temperate winter. 50–350 m. Limono-sandy, very little clayey.
  Middle atlantic Morocco (downstream part of the Souss plain).
- *Notocero bicorni-Calenduletum arvensis* Nègre (1956) 1959
  Thermomediterranean. Arid; temperate winter. Siliceous alluvium deposited on limestone.
  Middle atlantic Morocco (Sedd el-Messjoun, Hadra, Bahira, Bni-Amer).
- *Notocero bicorni-Spergularietum purpureae* Nègre (1956) 1959
  Thermomediterranean. Arid; temperate winter. Siliceous.
  Middle atlantic Morocco (Sedd el-Messjoun, Bahira north of El-Kelâa Sraghna).

Subassociations:

- *androcymbietosum graminae*
- *cladanthetosum arabici*
- *calenduletosum algeriensis*

- *Association with Aristida coerulescens* Br.-Bl. and Maire 1924
  Thermomediterranean. 500–600 m. Semi-arid; temperate to cool winter.
  Middle atlantic Morocco (jbel Zalagh).

Artemisietea vulgaris Lohmeyer, Preising and Tüxen ex von Rochow 1951

Perennial (sub)xerophilous ruderal vegetation.

Carthametalia lanati *S. Brullo in S. Brullo et Marceno 1985*

Onopordion castellani Br.-Bl. et O. de Bolos 1958 corr. Rivas-Mart. et al. 2001

(*Onopordion nervosi* Br.-Bl. et O. de Bolos 1958 corr. Rivas-Mart. 1975)

- *Chamaeleo gummiferi-Echietum boissieri* Müller, Deil, Galán de Mera and Vicente Orellana 2005
  Thermo- and Mesomediterranean. 0–1400 m. Sub-humid, Humid; warm, temperate winter. Limestone, clays, marls; deep soils.
  Steep slopes of road embankments and on grazed pastures.
  Tangerian peninsula and south Spain.
- *Notobasio syriacae-Scolymetum maculati* Rivas Goday ex Ladero, Socorro, Molero, M. Lopez, Zafra, Marin, Hurtado and Pérez-Raya 1981
  Thermomediterranean. Sub-humid; warm, temperate winter. Loam, clay, limestone.
  Rif (regions of Tangier and Dar Chaoui).

Silybo mariani-Urticion piluliferae Sissingh ex Br.-Bl. and O. de Bolos 1958

(*Silybion mariani* Rivas-Mart. in Rivas-Martinez et al. 1992)

- *Scolymo maculati-Silybetum mariani* Rivas-Mart. in Rivas-Martinez, Costa, Castroviejo and Valdés 1980

Thermomediterranean. Sub-humid; temperate, warm winter. Loamy.
Rif (region of Tangier).

Elytrigio repentis-Dittrichietalia viscosae *Mucina in Mucina et al. 2016*

Bromo-Oryzopsion miliaceae O. de Bolos 1970

- *Oryzopsio miliacei-Daucetum maximi* Bòlos and Vigo 1972
  Rif.

Chenopodietea Br.-Bl. in Braun-Blanquet, Roussine and Nègre 1952

(*Ruderali-Secalietea* Br.-Bl. et al. 1936 p.p.)
(*Geranio purpurei-Cardaminetea hirsutae* Rivas-Martinez et al. (1999) Rivas-Martinez et al. 2002)

Brometalia rubenti-tectorum *(Rivas Goday et Rivas-Mart. 1973) Rivas-Martinez et Izco 1977*

(*Thero-Brometalia annua* Rivas Goday et Rivas-Mart. ex Esteve 1973)

Cerintho majoris-Fedion cornucopiae Rivas-Mart. et Izco ex Peinado et al. 1986

- *Convolvulo cupaniani-Malopetum trifidae* Galán de Mera 1994
  Thermomediterranean. Sub-humid; warm to temperate winter. Marls.
  Rif (Tangerian peninsula).
  Subassociations:
  - *typicum*
  - *glossopappetosum macroti*

Echio plantaginei-Galactition tomentosae O. de Bolos and Molinier 1969

- *Trifolio pallidi-Vulpietum geniculatae* Galán de Mera 1995
  Thermo- and Mesomediterranean. Sub-humid, Humid; temperate, cool winter. Sandy or loamy soils.
  Rif.
- *Trifolium scabrum-[Echio-Galactition]-derivat community* Galan De Mera, Deil, Vicente Orellana and Müller 2004
  Thermomediterranean. Sub-humid; warm, temperate winter. Sands, marls.
  Rif (regions of Assilah and Tanger).

Hordeion murini Br.-Bl. in Br.-Bl. et al. 1936

(*Hordeion leporini* Br.-Bl. in Braun-Blanquet et al. 1936 corr. Bolòs 1962)

- *Anacyclo radiati-Hordeetum leporini* Bolos and Rivas-Martinez in Rivas-Martinez 1978
  Thermo- and Mesomediterranean. Sub-humid; temperate, warm winter. Various substrates.
  Northern atlantic Morocco (Gharb plain until Rabat) Rif (region of Tangier).
- *Hordeo leporini-Glossopapetum macroti* Peinado, Martinez-Parras and Bartolomé 1986
  Thermo- and Mesomediterranean. Sub-humid, Humid; temperate, warm winter.
  Rif.
- *Association with Malva parviflora and Hordeum leporinum* Br.-Bl. and Maire 1924
  Thermomediterranean. Arid; temperate winter.
  Middle atlantic Morocco (Haouz).

Taeniathero-Aegilopion geniculatae Rivas-Mart. and Izco 1977

- *Aegilops geniculata [Taeniathero-Aegilopion]-basal community* Galan De Mera, Deil, Vicente Orellana and Müller 2004
  Thermomediterranean. Sub-humid; temperate winter. Sandstone, clays.
  Rif (Tangerian peninsula: region of Melloussa etc.).
  Edges and slopes of roads and on the margins of small forests.

Chenopodietalia *Br.-Bl. in Braun-Blanquet et al. 1936*

Chenopodion muralis Br.-Bl. in Braun-Blanquet et al. 1936

- *Oxalo pedis-caprae-Fumarietum capriolatae* Poli 1966 ex Nezadal 1989
  Cited by Fennane (2003:101) with no details.
- *Association with Chenopodium murale* Br.-Bl. and Maire 1924
  Thermomediterranean. Arid; temperate winter.
  Middle atlantic Morocco (Haouz).
- *Urticetum membranaceae* Nègre 1959
  Thermomediterranean. Arid; temperate winter.
  Middle atlantic Morocco (Haouz)
- *Urticetum piluliferae* Nègre 1959
  Thermomediterranean. Arid; temperate winter.
  Middle atlantic Morocco (Haouz)

Other groupements

- *Malvetum sylvestris-subacaulis* Quézel 1952
  Oromediterranean. 3000–3400 m. Semi-arid; very cold. Feet of limestone cliffs.
  High-Atlas (Ghat massif).
  Nitrophilous association.
- *Association with Malva parviflora and Malva nicaeensis* Br.-Bl. and Maire 1924
  Thermomediterranean. Arid; temperate winter.
  Middle atlantic Morocco (Haouz).

Digitario sanguinalis-Eragrostietea minoris Mucina, Lososova and Silc in Mucina et al. 2016

Anthropogenic communities; thermophilous vegetation, rich in summer annual species.

Eragrostietalia *J. Tüxen ex Poli 1966*

Diplotaxion erucoidis Br.-Bl. in Braun-Blanquet et al. 1936

- *Oxalo cernuae-Cyperetum rotundii* El Antri 1983
  Thermomediterranean. Arid; temperate winter. Limestone, clayey.
  Citrus orchards and some industrial crops (summer beet, cotton) in Middle atlantic Morocco (Souss, Tadla).

- *Torulidetum arvenso-neglectae* El Antri 1983
  Thermomediterranean. Semi-arid; temperate winter.
  Mam and Man (fruit tree plantations).
- *Groupement with Brachiaria eruciformis* El Antri 1983
  Thermomediterranean. Semi-arid, Sub-humid; temperate winter.
  Northern atlantic Morocco (mainly in beet and cotton crops).

Gladiolo italici-Ridolfietalia segeti *Mucina* in *Mucina et al. 2016*

(*Secalietalia* Br.-Bl. 1931)
(*Chrysanthemetalia segeti* Nègre 1959)

Ridolfion segeti Nègre ex Rivas-Martinez et al. 1999

- *Capnophylletum peregrinae* El Antri and Montégut in El Antri 1983
  Thermomediterranean. Arid; Semi-arid; temperate winter. Clays.
  Middle atlantic Morocco.
- *Chrysanthemo segeti-Vaccarietum pyramidae* Nègre 1959
  Thermomediterranean. Arid; Semi-arid; temperate winter. Clayey, sandy.
  Middle atlantic Morocco (Bni-Moussa, Haouz).
- *Psoraleo americanae-Otospermetum glabri* El Antri 1983 em. Lastic 1989
  Thermomediterranean. Semi-arid, Sub-humid; temperate winter. Clayey, sometimes innonded.
  Northern atlantic Morocco (plain of the Gharb).
- *Triguero-Convolvuletum gharbensis* El Antri and Montégut in El Antri 1983
  Thermomediterranean. Semi-arid, Sub-humid; temperate winter. Clayey, little wet.
  Middle atlantic Morocco (Chaouia) Northern atlantic Morocco (Gharb).

Papaveretea rhoeadis S. Brullo et al. 2001

(*Stellarietea mediae* Tüxen et al. in Tüxen ex von Rochow 1951)
(*Ruderali-Secalietea* Br.-Bl. et al. 1936 p.p.)
Annual weed herbs; vegetation of arable crops, gardens, orchards...

Papaveretalia rhoeadis *Hüppe et Hofmeister ex Theurillat et al. 1995*

(*Centaureetalia cyani* Tüxen 1950 ex Von Rochow 1951)

Caucalidion lappulae Tüxen 1950 ex Von Rochow 1951

- *Cirsio acarnae-Isatisetum tinctoriae* (Nègre 1961) El Antri 1983
  Thermo-, Mesomediterranean. Sub-humid; cool, cold winter. Until 2000 m, rarely less than 500 m. Limestone stony, permeable.
  Moyen-Atlas (Aït Oufella, Col Tassount-ou-Filali).

Launaeion nudicaulis El Antri and Montégut in El Antri 1983

- *Psychino-Ceratocnemum rapistroidis* El Antri and Montégut in El Antri 1983
  Thermomediterranean. Arid; Semi-arid; warm, temperate winter. Dry, stony, loamy-sandy to clayey-sandy, limestones.
  Middle atlantic Morocco (Haouz, Souss).
  Grows mainly in barley crops.

Rumio-Brassicion barrelieri El Antri and Montégut in El Antri 1983

- *Ononido-Linarietum gharbensis* El Antri and Montégut in El Antri 1983
  Thermomediterranean. Semi-arid, Sub-humid; temperate winter. Sandy soils, more or less acidic.
  Middle atlantic Morocco.
  Messicole association.
- *Sclerosciadetum nodiflori* (Br.-Bl. and Maire) El Antri 1983
  Thermomediterranean. Arid, Semi-arid; warm, temperate winter. Sandy, dry, stony soils.
  Cereal crops in the Middle atlantic Morocco (Souss, Haha, Chiadma...)
- *Papavero rhoeas-Loeflingietum baeticae* Lastic 1989
  Thermomediterranean. Semi-arid, Sub-humid; temperate winter. Sandy.
  Northern atlantic Morocco (Maamora).

Sisymbrietea Gutte and Hilbig 1975

Zoo-anthropogenic plant communities, in and around animal shelters and disturbed ruderal habitats.

Sisymbrietalia Sophiae *J. Tx. ex Görs 1966*

(*Sisymbrietalia officinalis* J. Tüxen in Lohmeyer et al. 1962)

Geranio pusilli-Anthriscion caucalidis Rivas-Martinez 1978

- *Corydali heterocarpae-Succowietum balearicae* Daumas, Quézel, and Santa 1952
  Geographic repartition unknown; cited by Fennane (2003).
- *Geranio rotundifolii-Theligonetum cynocrambis* Rivas-Martinez 1978
  Geographic repartition unknown; cited by Fennane (2003).
- *Mercuriali ellipticae-Theligonetum cynocrambis* Peinado, Lorca, Martinez Parras 1982
  Thermomediterranean. Semi-arid; temperate winter.
  Mediterranean coast (Bokkoya) (Deil and Hammouni 1997).
- *Parietario mauritanicae-Ceratocapnetum heterocarpae* Martinez Parras 1982
  Thermomediterranean. Sub-humid; temperate winter. Rocky substrates.
  R (Tangerian peninsula).

Polygono arenastri-Poetea annuae Rivas-Mart. 1975 corr. Rivas-Martinez et al. 1991

Therophytic vegetation, rich with dwarf-herb in trampled habitats. Subcosmopolite.

Polygono arenastri-Poetalia annuae *Tüxen in Géhu et al. 1972 corr. Rivas-Martínez et al. 1991*

Polygono–Coronopodion Sissingh 1969

- *Poo annuae-Coronopodetum squamati* (Oberdorfer 1957) Gutte 1966
  Thermomediterranean. Sub-humid; temperate winter.
  Rif (Arba Ayacha).
  Subassociations:

– *hainardietosum cylindricae*

Polycarpion tetraphylli **Rivas-Martinez 1975**

- *Solivetum stoloniferae* Rivas-Martinez 1975
  Thermomediterranean. Sub-humid; temperate, warm winter.
  Rif (Tangerian peninsula).

Conclusion

The current syntaxonomic inventory of Morocco has nearly 670 associations (communities or groupements), spread over 97 alliances, 66 orders and 44 classes. These data reflect the richness and diversity of the country's natural ecosystems although they are weak, far from reflecting the reality on the ground. As a comparison, in the Iberian Peninsula, more than 2000 associations, 413 alliances, 141 orders and 76 classes were identified (Rivas-Martinez et al. 2002).

The gaps in phytosociological knowledge are still deep in Morocco. This discipline has been in decline for many years, while many communities, especially those dominated by herbaceous plants, have not yet been studied. Even worse, the situation presented in this work is certainly not the same on the ground, given the rapid degradation process suffered by most of our ecosystems which are not subject of any scientific monitoring.

Scientifically, as in practice, the phytosociology has never benefited from its rightful place in training establishments or in management plans. Maybe, it is a discipline difficult to approach, seems little practical, but it remains, until now, the best tool for describing plant communities, explaining their distribution, and handling aspects of exploitation, valorization, protection and conservation.

Finally, there is another aspect where phytosociology is very useful, even indispensable. We are thinking about habitat typology. This great project, unfortunately still very little addressed in Morocco, will not be possible without the development of syntaxonomic knowledge. It is time to do it; the present synthesis will certainly help to facilitate the task.

M. S. Taleb, M. Fennane, *Vascular Plant Communities of Morocco*, Geobotany Studies, https://doi.org/10.1007/978-3-319-93704-5

Acknowledgment

We are thankful to Professor Franco Pedrotti who initiated and supported this publication, and to our colleagues Ultrich Deil and Jean Paul Peltier for their interesting comments.

M. S. Taleb, M. Fennane, *Vascular Plant Communities of Morocco*, Geobotany Studies, https://doi.org/10.1007/978-3-319-93704-5

References

Achhal A (1986) Etude phytosociologique et dendrométrique des écosystèmes forestiers du bassin du N'fiss. Thèse doct d'état Univ Aix-Marseille III, Marseille, 240 pp + annexes

Achhal A, Akabli O, Barbero M, Benabid A, M'hirit O, Peyre C, Quézel P, Rivas-Martinez S (1980) A propos de la valeur bioclimatique et dynamique de quelques essences forestières du Maroc. Ecol Medit 5:211–249

Achhal A, Ziri R, Khattabi A (2004) Analyse phytosociologique des formations de matorrals du massif des Béni-Snassène (Maroc oriental). Acta Bot Malacitana 29:67–87

Aidoud-Lounis (1984) Contribution à la connaissance des groupements à sparte (Lygeum spartum L.) des Hauts Plateaux sud-oranais; étude phytoécologique et syntaxonomique. Thèse doct. 3ème cycle, Univ Alger

Varo Alcala J, Fernandez Casas J (1970) Notas sobre algunas comunidades béticas con Adiantum capillus-veneris. Ars Pharm 11:517–520

Allorge P (1922) Les associations végétales du Vexin français. Rev Gén Bot 33:113–115

Asensi A (1984) Limonietum emarginati (Crithmo-Limonion), nueva association para los sectors Gaditano y Tingitano. Doc Phytosoc NS 8:45–51

Barbéro M, Loisel R, Quézel P (1974) Problèmes posés par l'interprétation phytosociologique des Quercetea ilicis et des Quercetea pubescentis. CNRS Coll Intern 235:481–197

Barbéro M, Quézel P, Rivas-Martinez S (1981) Contribution à l'étude des groupements forestiers et préforestiers du Maroc. Phytocoenologia 9(3):311–412

Barbéro M, Benabid A, Quézel P, Rivas-Martinez S, Santos A (1982) Contribution à l'étude des Acacio-Arganietalia du Maroc sud-occidental. Doc Phytosoc NS 6:311–338

Baudière A, Simonneau P, Voelckel C (1976) Les sagnes de l'étang de Salses. Coll Phytosociol IV:35–42

Beeftink WG (1965) De zoutvegetatie van ZW-Nederland beschouwd in Europees vernands. Thesis. In: Wageningen

Beeftink WG, Gehu JM (1973) Prodrome des groupements végétaux d'Europe. Spartinetea maritimae I:1–43. Lehre

Benabid A (1982) Etudes phytoécologique, biogéographique et dynamique des associations et séries sylvatiques du Rif occidental (Maroc). Thèse Univ Aix Marseille. 199 p

Benabid A (1984) Etude phytoécologique des peuplements forestiers et préforestiers du Rif centro-occidental (Maroc). Trav Inst Sci Rabat Sér Bot 34:1–64

Benabid A (1988) Contribution à l'étude des groupements non forestiers du Moyen-Atlas central (Maroc). Bull Inst Sci Rabat 12:79–98

Benabid A (2000) Flore et écosystèmes du Maroc: evaluation et préservation de la biodiversité. Ibis Press Paris, 359 pp

M. S. Taleb, M. Fennane, *Vascular Plant Communities of Morocco*, Geobotany Studies, https://doi.org/10.1007/978-3-319-93704-5

Benabid A, Cuzin F (1997) Populations de dragonnier (Dracaena draco subsp. ajgal) au Maroc: valeurs taxinomique, biogéographique et phytosociologique. C R Acad Sci Paris, Sci Vie 320: 267–277

Benabid A, Fennane M (1994) Connaissances sur la végétation du Maroc: Phytogéographie, phytosociologie et séries de végétation. Lazaroa 14:21–97

Benabid A, Machrouh A, Schoenenberger A (1994) La végétation potentielle du Parc national du Souss Massa. Trav. coop. allemande GTZ. Administration des eaux & Forêts et de la Conservation des Sols, Rabat, 69 pp + 1 carte en 7 feuilles

Bendaanoun M (1991) Etude écologique de la végétation halophile, halohygrophile et hygrophile des estuaires, lagunes deltas et sebkhas du litttoral atlantique et méditérranéen et du domaine continental du Maroc. Thèse doct d'état Univ Aix-Marseille III, Marseille

Biondi B, Allegrezza M, Taffetani F, Wildpret W (1994) La vegetazione delle coste basse sabbiose delle isole di Fuerteventura e lazarote (Isole Canarie). Fitosociologia 27:107–121

Boer AC (1942) Plantensoziologische beschrijving van de orde der Phragmitetalia. Ned Kruidk Arch 52:237–302

Bolos O (1957) De vegetacione Valentina I. Collect Bot (Barcelona) 5(2):527–599

Bolos O, Vayreda A (1950) Vegetacion de las comarcas barcelonesas. Istituto Espanol de Estudios Mediterraneos, Barcelona

Borgesen F (1905) The algae vegetation of the Faeröese coasts, with remarks on the phytogeography. Botany of the Faeröes 2:683–734

Braun Blanquet J (1926) Etudes phytosociologiques en Auvergne. Rapport sur une excursion interuniversitaire. C. Mont St Louis, Clermont-Ferrand. 94 p

Braun Blanquet J (1930) Zentralalpen und Tatra, eine pflanzensoziologische Parallele. Verröff Ber Geobot Inst E T H Stifung Rübel 6:43 p

Braun-Blanquet J (1931) Aperçu des groupements végétaux du Bas Languedoc. Comm Sta Inst Géobot Médit Alp 9:35–40

Braun-Blanquet J (1933) Prodrome des groupements végétaux. Ammophiletalia et Salicornietea. Montpellier, 23 pp

Braun-Blanquet J (1936) Un joyau floristique et phytosociologique : "L'Isoecion" méditerranéen. Bull Soc Etud Sci Nat Nîmes 47:1–23

Braun-Blanquet J (1948) La végétation alpine des Pyrénées orientales. Consejo Superior de Investigaciones Cientificas, Barcelona

Braun-Blanquet J (1949) Überschist der Pflanzengesellschaften Rätiens III. Vegetatio 1:285–316

Braun-Blanquet J (1952) Prodrome des groupements végétaux de la France mediter-raneanne. CNRS. Service Carte group vég, Dir Carte group vég d' Afrique du Nord

Braun-Blanquet J, Bolos O (1958) Les groupements végétaux du bassin moyen de l'Ebre et leur dynamisme. Anales Estac Exp Aula Dei 5:1–266

Braun-Blanquet J, De Leeuw WC (1936) Vegetationsskizze von Ameland. Ned Kruidk Arch 46: 359–393

Braun-Blanquet J, Jenny H (1926) Vegetations-Entwicklung und Bodenbildung in der alpinen Stufe der Zentralpen. Denkschr Schweiz Naturf Ges 63:I–VIII, 183–349

Braun-Blanquet J, Maire R (1924) Études sur la végétation et la flore marocaines. Mém Soc Sci Nat Maroc 8(1ère partie):5–152

Braun-Blanquet J, Tüxen R (1952) Irische Pflanzengseeschaften. Veröff Geobot Inst Rübel Zürich 25:222–421

Braun-Blanquet J, Gajewski W, Wraber M, Walas J (1936) Prodrome des groupements végétaux. Fasc. 3. Groupemnts messicoles, culturaux et nitrophiles-rudéraux du cercle de végétation mediterranean. Comité Int Prodr Phytosoc, Montpellier, 37 pp

Braun-Blanquet J, Molinier R, Wagner H (1940) Prodrome des groupements végétaux. Cisto-Lavanduletea. Montpellier, 36 pp

Braun-Blanquet J, Roussine N, Nègre R (1952a) Les groupements végétaux de la France mediterraneanne. CNRS, Montpellier, 297 pp

Braun-Blanquet J, Pinto da Silva AR, Rozeira A, Fontes F (1952b) Résultats de deux excursions géobotaniques à travers le Portugal septentrional et moyen I. Une incursion dans la Serra da Estrela. Agron Lusit 14(4):303–323

Brullo S, Firnari F (1976) Les associazioni vegetali degli ambienti palustri costieri della Sicilia. Not Phytosoc 12:23–148
Brullo S, Marceno C (1985) Contributo alla conoscenza della vegetazione nitrofila della sicilia. Coll Phytosoc 12:23–148
Bueno A (1997) Flora y vegetacion dc los estuarianos. Cuadernos de Medio Ambiente, Naturaleza. Oviedo 3:1–334
Chapman VJ (1960) Salt marshes and salt deserts of the world, vol 392. Interscience Publishers Inc, New York
Christiansen W (1934) Das pflanzengeographische und soziologische Vergalten der Salzpflanzen mit besonderer Berücksichtigung von Schleswig-Holstein. Beitr Biol Pfl 22:139–154
Christiansen W (1955) Salicornietum. Mitt Florist-Soziol Arbeitsgem 5:64–65
Conard WS (1935) The plant associations of central Long Island. Am Midl Nat 16(4):503–511
Coquillard P (1982) Approche phytoécologique d'une enclave saharienne le Tafilalet. Thèse Univ Nice, 364 p + annexes
Corillion R (1953) Les halipèdes du Nord de la Bretagne: Etude phytosociologique et phytogéographique. Rev Gén Bot 60:609–775
Costa M, Mansanet J (1981) Les ecosistemas dunares Livantinos: La dehesa de la Albufera de Valencia. Anales Jard Bot Madrid 37(2):277–299
Dahl E Hadac E (1941) Strandgesellschaften der Insel Osty im Oslofjord. Saertr Nytt Mag Naturvidensk (Oslo) 82:251–312
Daumas P, Quézel P, Santa S (1952) Contribution à l'étude des groupements végétaux rupicoles d'Oranie. Bull Soc Hist Nat Afr Nord 43:186–202
de Bolos A, de Bolos O (1950) Vegetacion de las comarcas Barcelonesas. Instituto Espanol Estudios Mediterraneos, Barcelona
de Bolos O, Molinier R (1969) Vue d'ensemble de la végétation des îles Baléares. Vegetacio 17:251–270
de Bolos O (1970) Un groupement herbacée thermophile des Pyrénées centrales. Arch Natur Landschafts 10:139–143
de Bolos O, Viego J (1972) Contribution à la géobotanique de l'île de Majorque. Rapp Com Int Mer Médit 21(3):82–82
De Foucault B (1993) Remarques sur la végétation du Maroc et de l'Algérie. Bull Soc Bot Centre-Ouest, Nouv Ser 24:267–285
De Foucault B (2015) Contribution au prodrome des végétations de France: les Adiantetea capilliveneris Br.-Bl. ex Braun-Blanquet, Roussine & Nègre 1952. Acta Bot Gall 162:375–403
De Foucault B, Julve P (1991) Données phytosociologiques sur la dix-septième session de la Soc Bot Centre-Ouest. en Languedoc-Roussillon: réflexions sur les associa-tions arborescentes méditerranéennes. Bull Soc Bot Centre-Ouest 22:391–420
Deil U (1984) Zu Vegetation in Zentralen Rif (Nordmarokko) – unter besonderer Berücksich-tigung der Zedernwälder und ihrer Erssatzgesellschaften. Dissertationes Botanicae 74. Vaduz
Deil U (1989) Contribution à l'étude des groupements secondaires du Rif central (Maroc). Comparaison avec quelques séries de dégradation de la végétation ibérique. Stud Bot 8:79–90
Deil U (1994) Felsgesellschaften beiderseits der StraBe von Gibraltar. Hoppea, Denkschr Regensb Bot Ges 55(Honester-Festschrift):757–814
Deil U (1996) Adientea communities in the mediterranean area with general remarks about ecological scales within their habitats and about the saturation of plant. Phytocoenologia 26(4): 481–536
Deil U (1997) Vegetation coverand human impact. A comparaison of the Almarchal region (Gibraltar) and the Tangier hinterland (Morocco). Lagascalia 19:745–758
Deil U (2005) Distribution, Ecology and Phytosociology of the Rifean Endemism Perralderia paui. Hoppea, Denkschr Regensb Bot Ges 66 (Schönfelder-Festschrift):173–186
Deil U, Galan de Mera A (1996) Contribution à la connaissance de la phytosociologie et de la biogégraphie des groupements rupicoles calcaires du Maroc. Bull Inst Sci Rabat 20:87–111

Deil U, Hammoumi M (1997) Contribution à l'étude des groupements rupicoles de Bokkoya (littoral du Rif central, Maroc). Acta Bot Malacitana 22:131–146
Des Abbayes H, Corillion R (1949) Sur la répartition d'Ulex Gallii et d'Ulex nanus dans le massif armoricain. C R Somm Séances Soc Biogéog 228–229:86–89
El Antri M (1983) Approche synsystématique des groupements commensaux des cultures du Maroc. Coll Phytosc NS 12:283–311
Emberger L (1939) Aperçu général sur la végétation du Maroc. Verrof Geobot Inst Rübel Zürich 14:40–157
Esteve Chueca F, Fernandez-Casas J (1971) De vegetatione baetica, I. Cuad. Ci Biol 1:65–71
Fennane M (1982) Analyse phytogéographique et phytoécologique des tétraclinaies marocaines. Thèse Doct 3ème cycle. Univ Aix Marseille III, 146 pp
Fennane M (1988) Etude phytosociologique des tétraclinaies marocaines. Bull Inst Sci Rabat 12: 99–148
Fennane M (2003) Inventaire des communautés végétales à l'aide du phytosociologue au Maroc. Ecologia Mediterranea 29(1):87–106
Fernandez Casas J (1972) Notas fitosociologicas breves. Il Trab Dep Bot Univ Granada 1:21–57
Fernandez M, Santos A (1983) La vegetation de litoral de Canarias I. Arthrocnemetea. Lazaroa 5:143–155
Fukarek F (1961) Di vegetation des darss ihre Geschicht. Pflanzensoziologie 12:–321
Galán de Mera A (1994) Datos sobre la comunidades de Cerintho-Fedion en al SW de la Península Ibérica y NW de Africa. Nat Baetica 6:113–123
Galán de Mera A (1995) Trifolio pallidi-Vulpietum geniculatae nueva association del Echio-Galactition para el SW de la Peninsula Iberica. Doc Phytosoc 15:335–339
Galán de Mera A, Hagen de la Cerda MA, Vicente Orellana JA (1996) Variabilidad de las comunidades rupicolas brio-pteridofiticas en el SO de la península Ibérica y el NO de África. Orsis 11:7–13
Galán de Mera A, Vicente Orellana JA (1997) Sobre las comunidades con Poa bulbosa (Poaceae) del Medio Atlas central (Marruecos). Acta Bot Malacitana 22:227–230
Galan de Mera A, Deil U, Vicente Orellana JA, Müller JV (2004) Roadside vegetation in the campo de Gibraltar (SW Spain) and on the Tangier peninsula (NW Morocco). Stud Bot 23:63–93
Galán de Mera A, Sanchez Garcia I, Vicente Orellana JA (1997) Coastal plant communities of the southwestern Iberian Peninsula, Spain and Portugal. Phytocoenologia 27(3):313–352
Gehu JM (1976) Approche phytosociologique synthétique de la végétation des vases salées du littoral atlantique français (synsystématique et synchorologique). Doc Phytosoc 4:395–462
Gehu JM (1998) Schéma synsystématique des principales classes de végétation littorales sédimentaires européennes avec références à d'autres territoires holarctiques. Annali di Botanica LVI-1:1–52
Gehu JM, Biondi E (1996) Apport à la connaissance de la végétation du littoral marocain sud-occidental: les communautés végétales psammophiles des dunes et plages sableuses du Maroc macaronésien. Bull Soc Bot Centre-Ouest NS 27:179–214
Gehu JM, Biondi E (1998) Nature et limites de quelques végétations littorales de type macaronésien sur les côtes sud-occidentales du Maroc. Acta Bot Barc 45:439–453
Gehu JM, Sadki N (1995) Remarques de phytosociologie et de synchorologie comparées sur le littoral algérois. Doc Phytosoc NS 15:341–357
Gehu JM, Tüxen R (1975) Essai de synthèse phytosociologique des dunes atlantiques européennes. Coll Phytosoc 1:61–70 (La végétation des dunes. Paris 1971)
Gehu JM, Gehu J, Caron B (1978) Les groupements à Sedum anglicum des falaises nord-armoricaines. Coll Phytosoc 6:255–261
Guinochet M (1951) Contribution à l'étude phytosociologique du Sud Tunisien. Bull Soc Hist Nat Afr Nord 42:131–153
Gutte P (1966) Die Verbreitung einiger Ruderalpflanzengesellschaften in der weiteren Umbebung von Leipzig. Wiss Z Martin-Luther-Univ Halle-Wittenberg Math Naturwiss Reihe 15:937–1010
Harmsen GW (1936) Systematische Beobachtungen der nordwestdeutschen Seegrasformen. Ned Kruidk Arch 46:852–877
Hartog Den C, Segal S (1964) A new classification of the waterplant communities. Acta Bot Neerl 13:367–393

Hocquette M (1927) Étude sur la végétation et la flore du littoral de la mer du Nord de Nieuport à Sangatte. Arch Bot 1(4):1–179

Holub J, Hejny S, Moravec J, Neuhäusl R (1967) Übersicht der höheren vegetationseinheiten der Tschechoslowakei. Rozpr Cs Akad Ved, Praha, ser Math natr 77(3):1–75

Horvatic S (1934) Flora i vegetacija otoka Paga. [Flora and vegetation of the island of Pag]. Prirodoslovna istra zivanja kraljevine Jugoslavije 19:116–372

Horvatić S (1963) Vegetacijska karta otoka Paga s općim pregledom vegetacijskih jedinica Hrvatskog primorja. Prir Istraž, Acta Biol 4:3–187

Izco J, Fernandez Conzales F, Molina A (1984) El orden Tamaricetalia Br-Bl & Bolòs y su ampliacion con los tarayales hiperhalofilos. Doc Phytosoc NS 8:377–392

Kaabeche M (1996) La végétation steppique du Maghreb (Maroc, Algérie, Tunisie). Essai de synthèse phytosociologique par application des techniques numériques d'analyses. Doc Phytosoc NS 16:45–58

Klika J, Hadac E (1944) Rostlinna spolecenstva stredni Evropy [Plant communities of Central Europe]. Prıroda 36:249–259, 281–295

Klika J, Novac V (1941) Praktikum rostilinné sociologie, pudoznalstvi, klimatologie a ekologie. Praha

Koch W (1926) Die Vegetationseinheiten der Linthebene. Nordostschwiz Jb St Gall Naturwiss Ges 61:1–134

Ladero M, Socorro O, Molero Mesa J, López Guadalupe M, Zafra L, Marín G, Hur-Tado J, Raya FP (1981) Algunos consideraciones sobre las comunidades nitrófilas de Granada España. Anal Jard Bot Madrid 37:737–763

Lastic YP (1989) Les communautés adventices des cultures de la plaine du Gharb (Maroc). Dissertation, Fakultât fur biologie, chemie und geowissenschaften der Univ Bayreuth

Lemée G (1953) Contribution à la connaissance phytosociologique des confins saharo-marocains. Vegetacio 4:137–154

M'Hirit O (1982) Etude écologique et forestière des cédraies du Rif (Maroc). Thèse Doct Etat ès Sc Univ Aix Marseille III, 436 pp

M'Hirit O (1987) Etat actuel des connaissances sur la cédraie. Elément pour un programme de recherche. Comité CFA/CEF/CFPO Silva Mediterranea. Doc diffusion limitée, 38 pp + annexes

Maire R (1924) Etude sur la végétation et la flore du Grand-Atlas et du Moyen-Atlas marocains. Mém Soc Sc Nat Maroc VII:220 p

Martinez Parras JM (1982) Parietario mauritanicae-Ceratocapnetum heterocarpae ass nova. Anales Jard Bot Madrid 39:187–190

Martinez Parras JM, Peinado Lorca M (1990) Ensayo sobre la vegetacion rupicola basofila de la clase Asplenietea trichomanis en la provincia chorologica betica. Acta Bot Malacitana 15: 193–202

Médail F, Quézel P (1999) The phytosociological significance of SW Morocco compared to the Canary Islands. Plant Ecol 140:221–244

Meier H, Braun-Blanquet J (1934) Prodrome des groupements végétaux. Fasc. 2. Classe des Asplenietales rupestres groupements rupicoles. Mari-Lavit, Montpellier

Molina JA, Tahiri H, Agostinelli E, El Alaoui-Faris FE, Lumbreras A, Pardo C, Silva V, Pinto-Cruz C, Castoldi E, Navarro Campoamor J (2009) Contribución al conocimiento de la flora y vegetación de los humedales temporales del noroeste de Marruecos. Lazaroa 30:251–259

Molinier R (1934) Études phytosociologiques et écologiques en Provence occidentale. Ann Mus Hist Nat Marseille 27(1):1–273

Molinier R (1937) Etudes phytosociologiques et écologiques en Provence occidentale. Ann Mus His Nat Marseille XXVII:mém I

Mucina L et al (2016) Vegetation of Europe: hierarchical floristic classification system of vascular plant, bryophyte, lichen, and algal communities. Appl Veg Sci 19(Suppl. 1):3–264

Müller JV, Deil U, Galan de Mera A, Vicente Orellana JA (2005) Echium boissieri Steud. (Boraginaceae), an Ibero-Mauretanian roadside plant. Feddes Repert 116(3/4):226–242

Nègre R (1952) Les associations végétales du jbel Saa (Moyen-Atlas d'Itzer). Bull Soc Sci Nat Maroc 32:139–165
Nègre R (1956) Recherches phytosociologiques sur le Sedd el-Massjoun. Trav Inst Sci Chérif Rabat Sér Bot 10:1–193
Nègre R (1959) Recherches phytogéographiques sur l'étage de végétation mediter-raneanne aride au Maroc occidental. Trav Inst Sci Chérif Rabat Sér Bot 13:1–385
Nègre R (1960) Les associations végétales du Lac Zima. Bull Soc Sc Nat Phys Maroc 40:1–16
Nègre R (1961) Les associations végétales du jbel Saa (2ème partie). Bull Soc Sci Nat Phys Maroc 41:19–62
Nègre R (1977) Données phytosociologiques sur les associations thérophytiques du Maroc aride. Coll Phytosoc NS 6:23–32
Nezadal W (1989) Unkrautgesellschaften der Getreide und Frühjahrhackfruchtkulturen (Stellarietea mediae) im mediterranean iberian. Diss Bot 143:1–205. Stuttgart
Oberdorfer E (1952) Beitrag zur Kenntnis der Nordägäischen Küstenvegetation. Vegetatio 3:329–349
Oberdorfer E (1957) Süddeutsche Pflanzengesellschaften. Gustav Fischer, Jena
Oberdorfer E (1977) Süddeutsche Pflanzengesellschaften Teil I, 2nd edn. G Fischer, Stuttgart. 355 p
Passarge H (1964) Uber Pflanezngesellschaften der Moore in Lieberoser Endmoränengebiet. Abh Bericht Naturk Görlitz 39(1):1–26
Pawlowski B, Solowski M, Wallish K (1928) Die Pflaneznassoziationen des Tatra-Gebirtges VII. Teil Die Pflaneznassoziationen und die Flora des Morskie Oko-Tales. Bull Int Acad Polon Sci Cl Sci Math Ser B Sci Nat2: 205–272, + pl., carte
Peinado Lorca M, Martinez Parras JM (1982) Parietario mauritanicae-Ceratocapnetum heterocarpae ass nova. Anales Jard Bot Madrid 39:187–190
Peinado Lorca M, Martínez-Parras JM, Bartolomé C (1986) Notas sobre vegetación nitrófila II. Algunas novedades fitosociológicas en Andalucía. Stud Bot 5:53–69
Peltier JP (1982) La végétation du bassin versant de l'Oued Souss (Maroc). Thèse Doctorat d'Etat, Univ. Scientifique et Médicale de Grenoble, Grenoble
Peltier JP, Msanda F (1995) Diversité et originalité de la steppe à Euphorbia officinarum subsp. Echinus (Hook fil & Cosson) Vindt du sud-ouest marocain. Feddes Repert 106(3–4):215–229
Pérez Latorre AV, Galan de Mera A (1995) Datos sobre Rupicapnion africanae Br.-Bl. & Maire de la subregión Mediterránea Occidental. Acta Bot Malacitana 22:233–234
Pérez Latorre AV, Galan de Mera A (1997) Datos sobre Rupicapnion africanae Br.-Bl. & Maire de la subregión Mediterránea Occidental. Acta Bot Malacitana 22:233–234
Perez Latorre AV, Galan De Mera A, Deil U, Cabezudo B (1996) Fitogeografia y vegetacion del sector Aljibico Cadiz-Malaga, Espania. Acta Bot Malacitana 21:241–267
Pérez Latorre AV et al (2000) A new bryo-pterophytic epiphytic community from south-western Spain: Pterogonio gracilis-Davallietum canariensis. Crypt Bryol Lichenol 21:233–240
Pignatti S (1953) Introduzione allo studio fitosociologico della pianura veneta orientale con particolare riguardo alla vegetazione litoranea. Archiv Bot 29:1–25, 65–98, 129–174
Pignatti S (1954a) Introduzione allo studio fitosociologico della pianura veneta orientale. Arch Bot 12(28):265–329; 13(29/1):1–25; 13(29/2): 65–98; 13(29/3):129–174
Pignatti S (1954b) Introduzione allo studio fitosociologico della pianura veneta orientale. Atti Ist Bot Lab Critt Univ Pavia 5:92–258
Pons A, Quézel P (1955) Contribution à l'étude de la végétation des rochers maritimes du littoral de l'Algérie centrale et occidentale. Bull Soc Hist Nat Afr Nord 46:48–80
Preising E (1949) Nardo-Callunetea. Mitt Florist-Soziol Arbeitsgem N F 1:82–94
Quantin A (1935) L'évolution de la végétation à l'étage de la chênaie dans le Jura méridional. Comm Stat Int Géobot Médit Montpellier 37
Quézel P (1952) Contribution à l'étude phytogéographique et phytosociologique du Grand Atlas calcaire. Mém Soc Sci Nat Maroc 50:1–56
Quézel P (1953) Contribution à l'étude phytosociologique et géobotanique de la Sierra Nevada. Mem Soc Broteriana 9:5–77
Quézel P (1957) Peuplements végétaux des hautes montagnes de l'Afrique du Nord. Essai de synthèse biogéographique et phytosociologique. Edit Lechevalier, Paris, 463pp

Quézel P (1965) La végétation du Sahara du Tchad à la Mauritanie. Masson et Cie éditeurs, Paris

Quézel P, Barbéro M (1981) Contribution à l'étude des formations présteppiques à genévrier au Maroc. Bol Soc Brot Sér 2:53; 2:1137–1160

Quézel P Barbéro M (1986) Aperçu syntaxinomique sur la connaissance actuelle de la classe des Quercetea ilicis au Maroc. Ecol Médit 12(3–4):105–112

Quézel P, Barbéro M, Benabid A (1987) Contribution à l'étude des groupements forestiers et pré-forestiers du Haut-Atlas oriental (Maroc). Ecol Médit 8(1/2):107–117

Quézel P, Barbéro M, Benabid A, Loisel R, Rivas-Martinez S (1988) Contribution à l'étude des groupements préforestiers et des matorrals rifains. Ecol Médit 14(1/2):77–122

Quézel P, Barbéro M, Benabid A, Rivas-Martinez S (1992a) Contribution à l'étude des groupements forestiers et préforestiers du Maroc oriental. Stud Bot 10:57–90

Quézel P, Barbéro M, Benabid A, Loisel R, Rivas-Martinez S (1992b) Contribution à la connaissance des matorrals du Maroc oriental. Phytocoenologia 21(1/2):117–174

Quézel P, Barbéro M, Benabid A, Rivas-Martinez S (1994) Le passage de la végétation mediterraneanne à la végétation saharienne sur le revers méridional du Haut-Atlas oriental (Maroc). Phytocoenologia 22(4):537–582

Quézel P, Barbéro M, Benabid A, Rivas-Martinez S (1995) Les structures de végétations arborées à Acacia sur le revers méridional de l'Anti-Atlas et dans la vallée inférieur du Drâa (Maroc). Phytocoenologia 25(2):279–304

Rivas Goday S (1941) Contribución al estudio de la vegetación y flora de la provincia de Granada. Excursión botánica a Sierra Baza y Zújar

Rivas Goday S (1958) Nuevos ordenes y alianzas de Helianthemetea annuae Braun-Blanquet. Anal Jard Bot Madrid 15:539–651

Rivas Goday S (1964) Vegetacion y florula de la cuenca extremena del Guadiana. Publ Diputac Prov Badajoz, 777pp

Rivas Goday S (1970) Revision de las comunidades hispanas de la clase Isoeto-Nanojuncetea. Anales Inst Bot Cavanilles 27:225–276

Rivas Goday S, Galiano F (1959) Contribution al estudio de la *Quercetea hispanica*. Anal Inst Bot Cavanilles Madrid 17(2):285–406

Rivas Goday S, Ladero M (1970) Pastizales cespitosos de Poa bulbosa L., origen, sucesion y sistematica. Anal Real Acad Farm Madrid 36(2):139–181

Rivas Goday S, Rivas-Martinez S (1963) Estudio y clasificación de los pastizales españoles. Publ. Ministerio de Agricultura. Madrid

Rivas Goday S, Borja Carbonell J, Esteve Chueca F, Fernández-Galiano E, Rigual Magallón A, Rivas-Martínez S (1959) Contribution al estudio de la Quercetea ilicis hispanica. Anal Inst Bot Cavanilles Madrid 17(2):285–406

Rivas-Martinez S (1966) Esquema de la vegetation psamofila de las costas gaditanas. Albareda Herrera Barcelona:150–158

Rivas-Martinez S (1974) La végétation de la classe Quercetea ilicis en Espagne et Portugal. Anal Inst Bot Cavanilles 31:205–259

Rivas-Martinez S (1975) Mapa de la vegetacion de la provincia de Avilla. Anal Inst Bot Cavanilles 32:1493–1556

Rivas-Martinez S (1978) Sur la syntaxonomie des pelouses thérophytiques d'Europe occidentale. Coll Phytosoc 6:55–71

Rivas-Martinez S (1979) Brezales y jarales de Europa occidental (revision fitosocio-logica de las clases Calluno-Ulicetea y Cisto-Lavanduletea). Lazaroa 1:5–127

Rivas-Martinez S (1980) Estudio botanico de los ecosistemas de Donana (Huelva, Espana). Lazaroa 2:5–326

Rivas-Martinez S (1987) Memoria del mapa de series de vegetacion de Espana 1: 400.000. Ministerio de Agric, Pesca y Alimentacion, ICONA Madrid

Rivas-Martinez S (2011) Mapa de series, geoseries y geopermaseries de vegetacíon de España. Itinera Geobotanica NS 18(1+2):801pp

Rivas-Martinez S, Costa M (1976) Datos sobre la vegetacion halofila de La Mancha (Espana). Coll Phytosociol 4:81–97

Rivas-Martinez S, Costa M (1984) Sinopsis sintaxonomica de la clase Arthrocnemetea en la Peninsula Iberica. Doc Phytosoc NS 8:15–26

Rivas-Martinez S, Costa M, Castroviejo S, Valdés E (1980) La vegetacion de Donana (Huelva, Espana). Lazaroa 2:5–190

Rivas-Martinez S, Costa JC et al (1983) Sinopsis syntaxonomica de la classe Arthrocnemetea Braun Blanquet & Tüxen en la Peninsula Iberica. III Jornadas de Fitosociologia, la vegetation litoral Valencia, 26–28/9/1983.

Rivas Martinez S, Costa M, Itzco J (1986) Sintaxonomia de la clase Quercetea ilicis en El mediterraneo occidental. Not Fitosoc 19(II):71–98

Rivas-Martinez S, Lousa M, Diaz TE, Fernandez Gonzales F, Costa JC (1990) La vegetacion del sur Portugal (Sado, Alentejo y Algarve). Itinera Geobot 3:5–126

Rivas-Martinez S, Bascones J, Diaz TE, Fernandez Gonzales F, Loidi J (1991a) La vegetacion de Pirineo occidental y Navarra. Itinera Geobot 5:5–456

Rivas-Martinez S, Diaz TE, Fernandez Prieto JA, Loidi J, Penas A (1991b) Festuco hystricis-Ononidetea striatae y Rosmarinetea officinalis, clases de vegetacion independiantes. Itinera Geobot 5:505–516

Rivas-Martinez S, Costa M, Loidi J (1992) La vegetacion de las islas Ibiza Formentera (Islas Baleares). Itinera Geobot 6:99–235

Rivas-Martinez S, Fernandez Gonzalez F, Loidi J (1999) Checklist of plant communities of Iberian Peninsula, Balearic and Canary Islands to suballiance level. Itinera Geobot 13:353–451

Rivas-Martinez S, Diaz TE, Fernandez-Gonzales F, Izico J, Loidi J, Lousa M, Penas A (2002) Vascular plant communities of Spain and Portugal. Itinera Geobot 15(1) et (2)

Romo AM (1988) Drabo lutescentis-Violetum minimae ass. Nov. De Grand Atlas central. Acta Bot Barcin 37:331–334

Romo AM (2008) Contribution to the knowledge of rupicolous plant communities in the limestone areas of North Africa. Collect Bot (Barcelona) 27:75–90

Romo A (2009) Contribution to the knowledge of the Moroccan mountain vegetation. Collect Bot (Barcelona) 28:111–124

Santos Guerera A (1977) Ensayo sintaxonomico de la vegetacion de las Islas Canarias. La Laguna 205–220

Sauvage Ch (1963) Etages bioclimatiques. Notice et carte au 1/2.000.000, Atlas du Maroc, Sect II, pl 6b, Comité de géographie du Maroc

Scamoni A, Passarge H (1959) Gedanken zu einer natürlichen Ordnung der Waldgesellschaften. Archiv für Forstwesen 8:386–426

Sissingh G (1969) Über die systematische Gliederung von Trittpflanzengesellschaften. Mitt Florist-Soziol Arbeitsgem 14:179–192

Sunding P (1972) The vegetation of Gran Canaria. Skr Norke Vidensk – Akad Olso. Mat-Naturvidensk Kl NS 29:1–186 + 53 tab

Taleb MS, Fennane M (2003) Etude des groupements steppiques du Parc National du Haut-Atlas Oriental et ses bordures (Maroc). Bull Inst Sci Rabat Sci Vie 25:25–41

Taleb MS, Fennane M (2009) Etude phytosociologique des groupements présteppiques du Parc National du Haut-Atlas Oriental. Rev Écol (Terre Vie) 65:97–128

Tregubov V (1963) Etude des groupements végétaux du Maroc oriental mediterranean. Bull Mus Hist Nat Marseille 23:121–196

Tüxen R (1937) Die Pflanezngesellschaften Nordwestdeutschlands. Mitt Florist Soziol Arbeitsgem Niedersachsen 3:1–170

Tüxen R (1950) Grundriss einer Systematik der nitrophilen Unkrautgesellschaften in der Euro-sibirischen Region Europas. Mitt Flor Soz Arb Gem NF 2:94–175

Tüxen R (1960) Zur systematik der west und mitteliuropäischen Buchenzälder. Bull Inst Agron Stat Rech Gemblox HS 2:45–58

Tüxen R (1963) Kurze anmerkungen zur Exkursion des Internationalen vereinigeng für vegetationkunde. Vegetatio 11:395–400

Tüxen R, Oberdorfer E (1958) Die Pflanzenwelt Spaniens. II. Eurosiberische Phanerogamen-Gesellschaften Spaniens. Veröff Geobot Inst Rübel Zürich 32:1–328

Tüxen R, Preising E (1942) Grundbegriffe und methoden zum studium der wasser und sumpfpflanzengesellschaften. Dtsch Wasserwirtsch 37:10–17, 57–69

Vlieger J (1937) Aperçu sur les unités supérieures des Pays-Bas. Nederlandsch Kruidkundig Archief 47:335–353

Van Langendock HJ (1931) De vegetatie en oecologie der schorrenplanten Van Saafiongen. Bot Jaarbock Dodonae 23:1–128

von Rochow M (1951) Die Pflanzengesellschaften des Kaiserstuhls. Pflanzensoz 8:1–140

Westhoff V (1943) Plantensociologisch onderzoek, in het bijzonder op de Waddeneilanden. In: Handelingen van het XXIXe Nederlandsch Natuur-en Geneeskundig Congres, gehouden op 27, 28 en 29 April 1943 te Amsterdam, 27–40

Westhoff V, Dijk JW, Passchier H (1946) Overzicht der plantengemeenschappen in Nederland [Overview of the plant communities in the Netherlands]. B W Breughel, Amsterdam

Westhoff V, Van Leeuwen CG, Adriani MJ (1962) Enkele aspecten van vegetatie en bodem der duinen van Goeree, in het bijzonder de contactgordels tussen zout en zoet milieu [Some aspects of the vegetation and soil of the Coeree dunes, in particular of the contact zone between salty and fresh-water environments]. Jaarb Wetensch Genootsch Goeree-Overflakke 1961

Wildpret W, Del Arco M, Acebes JR (1983) Ononido-Cyperetum ass. nov. Vieraea 12(1–2): 349–357

Zeraia L (1981) Essai d'intréprétation des données écologiques, phénologiques et de production subéro-ligneuse dans les forêts de chêne liège de Provence cristalline et d'Algérie. Thèse doctorat d'état, Univ Aix Marseille III, 367 pp + annexes

Zobrist L (1935) Pflanzensoziologische und bodenkundliche Untersuchungen des Schoenetum nigricantis im nordschweizerischen Mitellande. Beitr Geobot Landesaufn Schweiz 18:1–144

MIX
Papier aus verantwortungsvollen Quellen
Paper from responsible sources
FSC® C105338

If you have any concerns about our products,
you can contact us on
ProductSafety@springernature.com

In case Publisher is established outside the EU,
the EU authorized representative is:
Springer Nature Customer Service Center GmbH
Europaplatz 3, 69115 Heidelberg, Germany

Printed by Libri Plureos GmbH
in Hamburg, Germany